The History of Yellow Fever

The History of Yellow Fever

An Essay on the Birth of Tropical Medicine

François Delaporte
foreword by Georges Canguilhem
translated by Arthur Goldhammer

The MIT Press
Cambridge, Massachusetts
London, England

This book was set in Sabon by Achorn Graphic Services Inc. and printed and bound by Halliday Lithograph in the United States of America.

Library of Congress Cataloging-in-Publication Data

Delaporte, François, 1941–
    [Histoire de la fièvre jaune. English]
    The history of yellow fever : an essay on the birth of tropical medicine / François Delaporte ; foreword by Georges Canguilhem ; translated by Arthur Goldhammer.
        p.   cm.
    Translation of: Histoire de la fièvre jaune.
    Includes bibliographical references.
    Includes index.
    ISBN 0-262-04112-X
    1. Yellow fever—History. 2. Tropical medicine—History.
I. Title.
    [DNLM: 1. Yellow Fever—history.   WC 530 D338h]
RC210.D4513   1991
616.9'28'009—dc20
DNLM/DLC
for Library of Congress   90-6086
                                                             CIP

to Georges Canguilhem

Contents

Foreword

François Delaporte's first published work, *Nature's Second Kingdom* (MIT Press, 1982), dealt with the history of plant physiology in the eighteenth century and in particular with vital questions of the day, the sexuality and reproduction of plants. The image of the flower bestowed on botany some of the flavor of pastoral poetry.

In his second book, *Disease and Civilization: The Cholera in Paris, 1832* (MIT Press, 1986), Delaporte turned his attention to a period in which the responsibility for an epidemic was divided between contagion and infection. No longer was he concerned with what Paul Verlaine in his *Romance sans paroles* apostrophized as "fruits, flowers, leaves, and branches."

This book introduces us to another phenomenon of the world of death. Insects, often associated in the imagination with flowers, are sometimes unwitting terrorists capable of sowing fear across entire continents. Yellow fever is carried by a mosquito. Delaporte writes: "Death came now not in the form of a man with a scythe but of a biting insect."

It took twenty years to prove decisively that the suspected agent was indeed responsible for the disease and that the Cuban physician Carlos Finlay was indeed its discoverer. The picture of the disease—complicated, as is often the case, by disputes over priority in which political interests cloaked themselves as love of truth—involves an orchestration of concepts elaborated by physicians and biologists during the bacteriological revolution. Today, in speaking of an epidemic disease, it seems quite simple to distinguish between the focus and specific agent, between the mode of transmission and diffusion, between the distribution of disease across populated territory and the composition of the organic environment. Yet the appropriate concepts of germ, vehicle, and intermediate host had to be laboriously elaborated by means of observation, analogy, experimentation, and refutation (some would say falsification). In the case of yellow fever great subtlety was required in analyzing the conditions resulting in infection. Properly speaking there is no transport unless the object transported is the same at the point of arrival as at the point of departure. If an alteration takes place along the way, what occurs is no longer transmission in the strict sense. The concept of *vector* was elaborated to combine two concepts that had proved useful in understanding other infections, that of vehicle and that of host. The *Culex* mosquito takes up the germ, harbors it during a period of incubation, and injects a product that it has in a sense fashioned.

The history of this conceptual elucidation, whose validity was demonstrated by its practical consequences for treatment and prevention, is also the history of various individuals, actors whose functions, work, and responsibilities involved them in the history—properly speaking, the political history—of the exploitation of the globe, of colonization, of international trade.

Since François Delaporte is one of a number of historians of science who do not make a profession of decrying epistemology, he has been able to arrive at an equitable determination of the respective contributions of each of the leading figures in this lengthy investigation: Manson, Finlay, Reed, and Ross. He has been able to avoid such errors as confusing a precursor with a founder, a word with a concept, a transport with a cycle of transmission. He has been able to discern, in this arduous and sometimes nasty controversy, an important biological advance, namely, the "complete redefinition of the alliances among living things." These alliances are sometimes deadly, just as in human societies. It is certainly true that the elucidation of yellow fever's mode of transmission altered the figure of Death. It became possible to trace on a map of the earth the boundaries within which Death has wings.

Georges Canguilhem

Acknowledgments

Georges Canguilhem aided me in this work with his encouragement and advice, for which I wish to thank him here. I have paid close attention to his views on methodology, as well as to those of Michel Foucault.

I also wish to thank Nathalie Gazeyeff and Santiago Ramirez Castaneda for their invaluable assistance. Thanks also to all those who aided me in my research: in Mexico, Roberto Moreno de los Arcos of the National Autonomous University of Mexico and Dominique Michelet and Jean Meyer of the Center for Mexican and Central American Studies; in Cuba, the Carlos Finlay Institute; in the United States, Gert Brieger and Caroline Hannaway of the Institute of the History of Medicine at Johns Hopkins University; and in France, Henri Mollaret of the Institut Pasteur.

The History of Yellow Fever

What is originality? *To see* something that as yet has no name, that cannot yet be named, even though it stands before everyone's eyes.

Nietzsche

Introduction

At once we encounter the inevitable *commentary*, seductive and misleading. The painter Valderrama's canvas records a historic scene: the moment when Finlay presented the eggs of the *Culex* mosquito to the American commission. Cornwell's painting depicts a similar scene: the production of the first case of yellow fever by American military physicians. The atmosphere of the two paintings is quite different, however. Valderrama portrays an interior scene. Finlay receives the American commission in his colonial residence at 110 Aguacate Street in old Havana. The discussion takes place around a table laden with the familiar accoutrements of the intellectual. In the manner of a naturalist Finlay is holding a porcelain cup that contains the mosquito's eggs. Cornwell depicts an outdoor scene. The setting is one mile outside the city of Quemados, where the American commission has established its medical station. In the center of the canvas is an operating table. In the manner of an experimental scientist Lazear has inverted a test tube containing a contaminated mosquito and placed it on Carroll's arm. The atmospheres of the two paintings are thus as different as a library is from a laboratory. The men gathered in Finlay's office are gathered around an object; those gathered in the improvised amphitheater are gathered around an experiment. Finlay's act is one of *exhibition*, Lazear's one of *demonstration*.

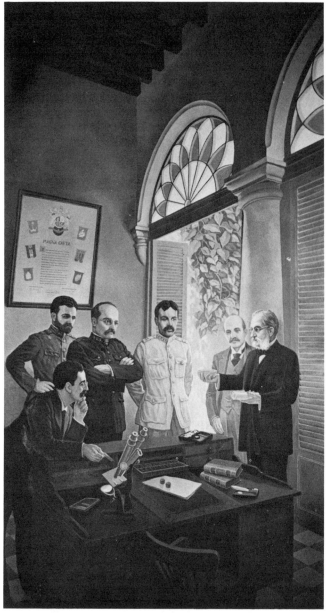

Esteban Valderrama, *Finlay entregando a la comison medica militar americana, huevos del mosquito trasmisor de la fiebre amarilla.* Courtesy of Museo Nacional, Havana (photo by Raidel Chao).

Dean Cornwell, *Conquerors of Yellow Fever,* third in the series "Pioneers of American Medicine."

The composition of the two paintings, symmetrical yet inverse, is striking. In Valderrama's painting the gazes converge on the porcelain cup, on the act of a naturalist which is also that of a teacher. The American commission forms a circle around Finlay. It receives from the Cuban physician not only an object but also something in the nature of a lesson. Not without misgivings: Reed, his hands behind his back, is polite but reserved. Carroll, his arms crossed, is frankly skeptical. Only Lazear, hands on hips, displays a genuine interest. How could he not be drawn toward what Finlay is saying? He was one of the few physicians in the United States with a thorough knowledge of malaria. Had some future experiments already begun to take shape in his imagination? And we must not forget the witnesses to the scene: with Finlay are his old friend Dr. Diaz Albertini and, leaning on the desk, his son. We may wonder, as they

4   did, how the interview would turn out. For the situation
was tense. But Finlay's gesture—which may strike some as
decisive, others as supplicating—leaves no doubt about the
outcome of the visit: the commission will leave with the
precious gift.

In Cornwell's painting roles and attitudes are reversed. Lazear is inoculating Carroll, the subject of the
experiment. On either side of the table, looking on
expectantly, are Wood, the governor-general, and Kean,
the chief physician. Reed, standing, seems fairly certain of
the result. He is in charge of the operation. The master of
ceremonies, he seems already to be posing for posterity. In
the background, watching, is the squad of brave volunteers, on this occasion in the role of attentive students.
Finlay, along with Agramonte and Truby who are just
behind him, are simply spectators. Visibly anxious, he
places his left hand on his heart, no doubt to contain his
emotion. The moment has come for Finlay to show that he
was a Finlayan, and right to have been one all along.

The two scenes are separated by barely three weeks.
The visit to Finlay took place on August 1, 1900, and
Lazear inoculated Carroll on August 24. Note, incidentally, that the suits and uniforms are of the period. Not a
gaiter or button is missing. The errors are of a different sort.
In Valderrama's painting Agramonte, the fourth member
of the commission, is absent. In Cornwell's the inaccuracies are even more flagrant: Reed, Finlay, and the
others were not in fact present at the scene. The inoculation, more prosaically, took place without witnesses in a
hospital room at Camp Columbia. Yet it ill behooves us to
speak of errors and inaccuracies. Both paintings are obviously allegorical. Painted in response to official commissions, symbolic values were involved. Valderrama's
painting glorifies Cuban medicine; Cornwell's does the
same for American medicine. The absence of Agramonte
(who was Cuban) in Valderrama's painting allowed the

painter to bring out a nice contrast: Finlay confronting the three brilliant American army doctors. Cornwell's recall of everyone involved in the story is a striking device: their presence at the time of a crucial experiment was dramatically necessary.

The two paintings succeed in what they set out to do. They record two important moments in a memorable adventure fraught with consequences: soon the scourge would be eradicated. The two paintings are remarkable: they attach a name to each of these two moments. Finlay is credited with the conception of the theory and the discovery of the correct species of mosquito. Reed is credited with the exemplary demonstration of the mosquito's role in the spread of the disease. The museum-goer will learn that Finlay hazarded a theory and that Reed risked the life of Carroll to verify it; he will learn the names of mankind's two benefactors. Most histories of yellow fever research have been content to set forth these two scenes, which offer an example of scientific collaboration and suggest an equitable distribution of credit. In order to account for the lapse of time between Finlay's formulation of his hypothesis and its resurrection by Reed, it was enough to conjure up certain imaginary obstacles. The sociology of science to the rescue: imperialism's pact with medical science was not yet sealed.

There were also polemical commentaries, however. Underlying the two paintings were two divergent interpretations, whose differences could be brought out. One might choose to emphasize Finlay's pioneering work. This version of history is the one generally preferred by Cuban historians. Finlay, cast as the founder of a new school of thought, plays the lead, while the American commission is reduced to a subordinate role: it confirmed a theory it did not conceive using a species it did not discover. But one may also choose to place the accent on the confirmation as the great achievement. This version is the one generally preferred by American historians. Reed gets the leading role, while Finlay is relegated to the status of visionary. He merely repeated, we are told, what others had said before him, and the mosquito he captured was of the most common species. These antagonistic versions are based on a common principle. Specific individuals are credited with originality for their work in formulating or verifying a theory. Everything begins and ends with Finlay, or else with Reed. Everything begins and ends with the genius of the founding saints.

Thus we have conciliatory accounts and antagonistic accounts of the event. And it cannot even be said that this naive scientific iconography was responsible for limiting the views of historians, for these paintings were in fact inspired by their books.

I envision a different sort of iconographic analysis. Its purpose will be to bring out those objects, concepts, and theories implicit in the description but not explicitly mentioned. If nothing else, this approach will at least avoid the predictable. The two paintings are interesting, then, not so much for what they show as for what they hide. I see the presentation of the *Culex* eggs to the American commission as an admission of *failure*. I see the experimental inoculation as an indisputable *success*, but one whose glory indubitably redounds to Ross. Before "conquering" yellow fever, was it not necessary to succumb to the appeal of his ideas? As for Finlay, we know that he attempted more than a hundred times to place a test tube containing a contaminated mosquito on the arm of a selected volunteer—without success. If similar attempts sometimes succeeded and sometimes failed, we may suspect that the experimental protocol was not the same. The same act may then carry different meanings. The rest would follow. If Finlay and Lazear were both dealing with the same mosquito, they conceived of that mosquito in different ways: the object was the same, but it was not the same *research object*. Nor was the same *hypothesis* involved: an *agent of transmission* differs from an *intermediate host* as a mechanical device differs from a biological process.

A number of questions follow from this analysis. Why did Finlay keep the *Culex* to himself for twenty years before making a gift of it to the American commission? How did the commission arrange the concrete historical conditions that ensured the success of its mission? Why was it not until the end of the century that anyone thought of the modest Cuban doctor?

In order to answer these questions, we must ask ourselves what a painter commissioned to do a canvas in honor of British tropical medicine might have painted. To answer this question requires no great effort of the imagination. Two scenes come immediately to mind. The first is set in London in 1894. Manson receives Ross in the small laboratory he has built in the attic of his apartment at 21 Queen Anne Street in Cavendish Square. Ross is shown bending over a microscope. For the first time he observes the *Plasmodium*. Manson is shown exhibiting a specimen of the *Culex* mosquito. The scene records the historical moment when Manson set Ross on the trail of the propagative cycle of the malarial agent by showing him Laveran's discovery and by suggesting to him his working hypothesis: his theory, elaborated some time between 1877 and 1880, concerning the mosquito's role in the transmission of filariasis. The second scene takes place four years later, in 1898. Ross is flanked by two cages. One contains sparrows that have died after being exposed to mosquitoes that have fed on a sparrow whose blood contained parasites. The other contains healthy sparrows that have been protected by a mosquito net. This painting would thus depict the historical moment when Ross produced the first experimental infection. The scene of the painting is not Ross's tiny laboratory in Calcutta. Instead, the (imaginary) painter would have captured his subject at the place where he first heard the good word from Manson: at Edinburgh in the hall where the British Medical Association has just met. Among the audience would surely be Chamberlain, Lord Lister, Osler, and Manson.

These two hypothetical paintings are oddly similar to the two paintings described previously. Just as Manson put Ross on the path leading to the discovery of malaria's mode of transmission, so Finlay put Reed on the path leading to yellow fever's mode of infection. In both cases, moreover, the end of the story involves the same spectacular reversal of fortune: just as Ross succeeded where Manson failed, so Reed succeeded where Finlay failed.

Make no mistake: there is no parallel between Manson and Finlay or between Ross and Reed. There are, however, quite real *affinities*, which can help us to answer our questions. Why did Finlay trap the *Culex* mosquito? Finlay was indebted to Manson for an object, the mosquito, from which he formed the notion of the insect as agent of transmission. What accounts for the success of the American commission? It was indebted to Ross for the hypothesis that the mosquito serves as intermediate host. It therefore paid close attention to the period during which the yellow fever germ incubates in the insect's body and produced an experimental protocol that was the key to producing a successful experimental case. Why did Finlay's hypothesis remain in scientific limbo for twenty years? It took that long to unravel the mechanism of malarial infection. The resolution of that mystery yielded two seemingly contradictory effects: Ross's work rescued Finlay's hypothesis from oblivion but simultaneously rendered it irrelevant. Finlay was right to have put his finger on the pertinent relation and to have designated the correct species of mosquito, but his work was irrelevant insofar as the hypothesis of a mechanical transport of the infection was false.

It is immediately clear how dear a price historians have paid for their lack of attention to the work of Manson and Ross. Unwittingly they erected an insurmountable obstacle to understanding the history of research on yellow fever.

Among other things this book deals with parasitology, entomology, and epidemiology. It treats the discovery of the importance of anthropods in tropical medicine. In the last two decades of the nineteenth century a new domain of objects, concepts, and theories emerged. We must analyze the historical conditions that made this new realm of knowledge possible and explore its experimental limits and rational structure. Research on filariasis, malaria, and yellow fever produced a set of interelated discourses. We must first describe how these related domains of inquiry overlapped. We must then explore the limits and intersections of the several discourses in question. Finally, we must clarify the exchange of methods, techniques, and models among workers in these various areas.

It is tempting to begin by describing what was known about yellow fever in the 1880s, but that temptation must be firmly resisted. An inventory of the theories that could be found in any treatise on tropical pathology would tell us very little. Such a treatise would have been obsolete even before its publication. The whole field was in the process of being transformed by the work of physicians scattered throughout the Caribbean or practicing in China. I have therefore chosen instead to examine what Finlay was reading. There are two advantages to this approach. First, it frees me from the obligation to be exhaustive; I need only note the works used by the Cuban physician. Second, I am assured of including those texts that proved decisive in the sense that they actually guided Finlay in his research.

In order to say what previous work influenced Finlay, we must state by what criteria we define influence. What propositions, already formulated in other works, became fundamental elements of Finlay's own discourse? One might argue that the important propositions were those that Finlay accepted as necessary assumptions or models, while other propositions, which he repeated as mere historical curiosities, may safely be neglected. Such an approach is legitimate and illuminating provided that two conditions are met: the reconstruction of his reading must be *complete* and his judgment as to the status of any particular proposition must be immediately *legible*. In fact, however, Finlay was anything but candid about his reading and the uses to which he put it. He was discreet to the point of dissimulation. It would be a waste of time to track down the references he omitted or the sources for statements whose position in the overall architecture of his work makes it impossible to identify their origin. Nor can we ask about his attitude toward the work of his predecessors; it is hidden. In short, our knowledge of the influences that shaped his thinking is *incomplete* and his judgment of earlier work is *indecipherable*. Thus an empirical approach is out of the question.

How, then, can we bring to light what remains hidden, identify the influences on Finlay's work, and describe how he used the work of his predecessors? I propose to do so by focusing on the problems that Finlay considered. In this way we shall be able to identify what works Finlay must have read in order to pose, and subsequently resolve, those problems. My approach will be guided by identification of this set of interrelated problems (*problématique*).

Three avenues of research stand out. The first concerned the problem of transformation of the yellow fever germ outside the human body. The second pursued the mode of propagation of the disease. Both ended in failure. Finlay in fact borrowed solutions to these two problems from Manson and then showed that those solutions were incorrect. These failures proved useful, however, for they led the Cuban physician to consider the mosquito as an agent of transmission and thus opened up a third avenue of research that ultimately led to formulation of the hypothesis with which we are concerned here. In order to pursue this avenue of inquiry, Finlay had to come to grips with three leading studies of the day, studies that defined the scientific context outside which Finlay's work would have been impossible. They were the *Report on Yellow Fever in the U.S.S. Plymouth in 1878–'9* (1880), Manson's research on *Filaria bancrofti* (1877–80), and the "Preliminary Report of the Havana Yellow-Fever Commission" (1879). These documents are necessary and sufficient to reveal the path by which Finlay arrived at his theory of 1881. Let us begin, then, with a brief discussion of the several strands that Finlay's work combined.

THE PLYMOUTH REPORT

On February 25, 1880, the Department of the Navy appointed a commission to look into the cause of an outbreak of yellow fever on board the *U.S.S. Plymouth*. The investigation was to be painstaking. The commission was to report, first of all, on the circumstances in which the vessel found itself in the course of its various cruises. It was then to inspect those parts of the ship likely to harbor an infection. Finally, it was to determine whether the second outbreak of the disease, in March 1879, was related to the first outbreak in November 1878. The final report was to include a medical history of the ship.

What events lay behind the instructions issued by Dr. Wales, the Navy's chief physician? On October 6, 1878, the *Plymouth*, a United States Navy steamer, left Portsmouth for St. Thomas, where it arrived on October 19. In St. Thomas the vessel replenished its supply of coal. The crew was confined to the ship, but the officers and employees of the supply company were free to go ashore. On October 25 the ship left St. Thomas for Santa Cruz, where it remained anchored a half-mile offshore until November 7. Once again the crew was confined to the ship, but the officers and supply personnel were free to come and go. The weather was hot and humid. On

November 5 seven cases of yellow fever were diagnosed on board ship. The sick men were immediately transferred to the Santa Cruz hospital. The ship resumed its course and arrived in Boston on December 17. There it was exposed to the harsh winter cold and subjected to fumigations with sulfuric acid. On March 15, 1879, the ship sailed for the Antilles. On March 22 two cases of yellow fever broke out. The *Plymouth* immediately turned north. The epidemic disappeared. On April 7 the ship arrived in Portsmouth and was placed in quarantine.

Three months later, medical inspector Dean submitted his report. Here is its conclusion:

In the first place, we find lying at the very root of the evil described the faulty construction, in a sanitary sense, of the ship herself. . . . In such a ship the sound condition of the vessel and of the crew she carries are alike imperiled.

The second consideration which presents itself is that she was provided with stores brought from a place long known to be dangerously infected with the poison of yellow fever. Many of the packages examined by the board bore the mark of the naval store house on the isle of Enchados in the harbor of Rio de Janeiro, where many fatal cases of yellow fever have been engendered. Some of these packages were of a nature well calculated to hold and carry the germs of the disease even through a cold climate.

The third point to be noted is that just before the appearance of the fever on board, the *Plymouth* had taken a great quantity of coal from St. Thomas, a dreaded source of yellow fever, and where the disease was prevailing at the time.

From one or both of these causes of infection came the germ or seed which found a most congenial soil for its fructification in the decayed ship, which had been cruising for years in the yellow fever zone, and which contained beneath the flooring of her holds and store-rooms the putrefying materials which are almost invariably found associated with outbreaks of this disease.

Under these highly favorable conditions the fever naturally became epidemic, as the flame follows the spark falling among inflammable materials.[1]

Here we have an excellent example of knowledge of yellow fever, such as it was in 1880, applied to a concrete situation. The Plymouth Report offers a snapshot of contemporary medicine, certain traits of which are accentuated better than any abstract exposition of medical theory could do. By forcing medical thought to confront a problem of this kind, the outbreak of yellow fever tested medicine's possibilities and limits. Hence it is best to consider medical thought about the question in the United States, where it was at its most vigorous. The preeminence of American medicine was recognized at the time in Cuba, if not in Europe.

The conclusion of the Plymouth Report emphasizes certain key ideas. The etiology of the disease is envisioned in terms of parasitic theory. In the references to a "germ or seed" of the disease and to the "poison of yellow fever" there is clear allusion to some kind of microorganism or, more accurately, to a transportable pathogenic agent capable of attaching itself to various objects. It is acknowledged that yellow fever can exist in either an endemic or an epidemic state. Regardless of the form it takes, however, the places in which it appears are regarded as sources of infection, meaning not just receptacles of germs but, even more, environments favorable to the germ's development. The conditions required for the germ to flourish and reproduce are partially known: a hot, humid, unhealthy climate. Thus the theory incorporates the prevailing belief in parasitic doctrine, establishes the malady's mode of propagation, and raises the question of the type of environment in which the germs are most likely to revive.

It was assumed at the time that the pathogenic agent of yellow fever was a specific germ. This hypothesis was based on the observed sequence and regularity of pathological phenomena, which argued in favor of a unity of cause. This clear-cut specific etiology was further reinforced by two complementary observations: immunity, in the sense that a single attack appeared to protect an individual against further infection, and transportability of the pathogenic agent. Yellow fever was classed among the parasitic diseases: its agent was *"specific,* that is, it reproduces yellow fever and no other disease. . . . The necessary inference from this fact is that the cause is always the same, and that it reproduces its like propagated, a peculiarity of living organisms."[2]

The remaining question, then, was how the disease was propagated. A disease may be considered contagious when the virus adheres to the body of the afflicted individual and when its presence is immediately reflected in symptoms. It was noticed that the persons most exposed to the disease, such as those administering treatment, did not contract it. It was also known that attempts to produce contamination by artificial means using supposedly germ-bearing objects had resulted in failure. There was no need to demonstrate that pathological effects were slow to declare themselves. Repeated experience proved that a considerable period of time elapsed between a ship's exposure in an infected port and outbreak of the disease on board. Contagion was therefore ruled out, and yellow fever was classified as an infectious disease. "The essential poison of yellow fever is not a product of disease in the human subject, like the poison of small-pox, for example, but is produced and developed outside of the body."[3]

Observation of epidemics yielded several significant findings: because the malady seemed to flourish in unhealthy conditions and hot, damp climates, the yellow fever germ was said to be a plantlike organism. The epidemic was frequently associated with unhealthy shipboard conditions. Under such conditions, it was observed, mushrooms often grew on rotten wood. This was taken as a sign: just as shipboard conditions favored the growth of lower plant species, so, too, did the same conditions favor the growth of the yellow fever germ. Similarly, the epidemic was associated with hot, humid climes, and again the behavior of plants was interpreted symbolically: just as plant life could withstand the cold in the form of spores, so, too, could the yellow fever germs remain dormant "so long as the surrounding conditions [were] unfavorable to their developments."[4] Once conditions turned favorable, the germs again became noxious and instigated the disease.

Many peculiarities of yellow fever could be explained by examining the behavior of the germ in the light of what was known about the natural history of the lower plant forms. With the aid of this analogy it immediately became clear how the germs could be taken aboard ship in a dormant state and remain in that state until conditions favored their revival, whether owing to the storage of putrid cargo or to the ship's return to a warm latitude. This was what happened to the *Plymouth*. "The poison manifests itself under certain conditions, such as are favorable to the growth of organic life, namely a temperature above 72 degrees Fah., moisture, and the presence of decomposing matter."[5]

It soon became apparent, however, that these factors were not sufficient to revive the infectious principle. For one thing, unhealthy conditions did not always lead to an outbreak of the disease. On long voyages to India and China not even the most unsanitary shipboard conditions had ever given rise to yellow fever. In many tropical regions, moreover, the disease was unknown. What is more, even in countries where yellow fever was endemic, the disease did not always declare itself. Periods during which the affliction was prevalent alternated with periods during which it was rare, as was evident from the tables Béranger-Féraud had compiled in Martinique: the longest period of remission was nine years, and the longest period of epidemic was thirteen years. La Roche, moreover, had published an important set of observations showing that appropriate levels of temperature and humidity did not always yield an "epidemic constitution."

Most physicians had noticed this strange characteristic of yellow fever and therefore envisioned an additional factor. If the yellow fever germs were not always active when heat and humidity levels were suitably high, "an undetermined factor, presumed to be meteorological," must also be at work.[6] This hidden variable must be implicit in the environment, an element that stands in relation to the germs as the spark stands to the powder. In short, the outbreak of an epidemic depended on the conjunction of several factors, the most important of which remained unknown.

Still, the hypothesis that the germs are transformed in the environment was not incompatible with the idea that they originate with the afflicted individual. It was possible, without contradiction, to accept the theory of the *nidus* while at the same time reactivating the idea that some human filiation must be involved. Admittedly, that filiation had to be indirect, because between the time the germs left the patient and the time they were received by the healthy individual the transformative power of the *nidus* had to come into play.

Most American doctors rejected the idea that the germs come from the patient. They did so because they believed that such a provenance was possible only with contagious diseases. Since yellow fever did not appear to be a contagious disease, it seemed clear that the germ must reside initially in the environment. "Dr. Bemiss is one of the few observers of great experience and high authority who hold a contrary opinion. 'The poison is reproduced chiefly, if not wholly, within the body, but,' as he goes on to say, 'undergoes some change after its escape from the body, which increases its toxic qualities.' "[7] This was not a new idea. At the time Budd and Pettenkofer regarded typhoid fever and cholera as indirectly transmissible diseases. They assumed that specific germs emitted by an afflicted individual must be transformed outside the body before exerting a morbid effect on another person.

Bemiss therefore saw yellow fever as a disease analogous to other indirectly transmissible diseases. He encountered the same difficulty as his colleagues, however. What was the mysterious factor that determined the toxicity of the germ? The hypothesis of a plantlike poison led to exploration of the environment: some invoked a "generative atmospheric disposition" (Béranger-Féraud); others "solar radiations" (Barlow); and still others "telluric conditions," "atmospheric pressure," or even some kind of "pabulum" (Bemiss). In fact, no physicians were under any illusion about the value of such explanations. All very likely subscribed to the notion that "the discovery of this unknown factor in the generation of yellow fever epidemics would be a great boon to humanity."[8]

As far as the mode of propagation was concerned, Bemiss was not in disagreement with those who saw yellow fever as a purely infectious disease and who spoke of an "invasion of the poison by the way of the atmosphere."[9] Bemiss, who saw yellow fever as an indirectly transmissible disease, was free to choose between a digestive and a respiratory pathway. The epidemiology of yellow fever was not identical with that of cholera or typhoid. A digestive pathway could be ruled out: "There is no fact connected with the epidemic of 1878 which justifies a belief that yellow fever infection is ever received through food or drinks." That left the respiratory pathway: "The disease is usually contracted by inhaling infected air. Yellow fever must be numbered among those epidemic maladies whose special poisons are air-borne, and through that medium distributed."[10]

Manson to T. Spencer Cobbold (Amoy, 20 June 1879):

I will forward you by this mail filaria-impregnated mosquitoes. They are preserved in glycerine and were fed on the blood of the man whose case I append. . . . I therefore determined to send you the first scrotum I amputated in which I had unquestionable evidence of filariae. . . . The case is of much interest as it exhibits, first, the transition from lymph-scrotum to elephantiasis; second, it demonstrates unmistakably that the parent worm is not necessarily present in the affected tissues themselves. . . . Third it illustrates well a new fact in the history of the filaria. The young escape into the circulation at regular intervals of twenty-four hours, the discharge commencing soon after sunset and continuing till near midnight, from which time until the following noon their numbers gradually decrease. . . . It is marvelous how nature has adapted the habits of the filariae to those of the mosquito. The embryos are in the blood just at the time the mosquito selects for feeding. . . . When the mosquito penetrates a blood vessel, the passing embryos, lashing about as is their habit, entangle themselves on the proboscis and get sucked up. Hence the enormous number of embryos in the mosquito's stomach and the selecting faculty of that insect.

It seems to me that Lewis by his great discovery has opened a new field in tropical pathology. The interest and importance of *F. bancrofti* and *Filaria sanguinis hominis* is by no means exhausted yet. I hear Lewis is in England, but when he returns to India I hope he will take up the subject again.[11]

In 1879 Manson was already in a position to summarize the history of research on filariasis. Lewis had discovered the embryonic form of the nematode and Joseph Bancroft the adult form. It was Manson himself, however, who first described the metamorphoses of the tiny worm and determined part of its life cycle. We must pause briefly to examine the significance, extent, and limits of the Amoy physician's work.

In 1872 Lewis showed that the microscopic hematozoon that Wucherer had found in the urine of a patient suffering from chyluria was also present in the patient's blood, whence the name *Filaria sanguinis hominis*. Shortly thereafter, in Brisbane, Joseph Bancroft found the adult form of the embryo in a lymphatic abcess. He communicated his discovery to Cobbold, who published it in July 1877 and who named the microfilaria's parent *Filaria bancrofti*. In 1877 it was therefore known that the tiny nematodes found in the blood and urine were the offspring of an adult worm lodged in the lymph vessels. But how was the transition from embryo to adult accomplished? How did the mocrofilariae exit the human circulatory system? By what avenue were they reintroduced, at a more advanced stage of development, into another individual? During 1877 and 1878 Manson would answer these questions. His theory had by then been formulated and was known to the scientific community. In March 1878 Cobbold was able to tell the Linnean Society: "The character of the changes undergone by the microscopic *Filariae*, and the ultimate form assumed by the larvae whilst still within the body of the intermediate host (*Culex mosquito*) are amply sufficient to establish the genetic relationship as between the embryonal *Filaria sanguinis hominis*, the stomachal *Filariae* of the mosquito, and the sexually mature *Filaria bancrofti*."[12]

It was Manson who opened up a new field in tropical pathology. Note how he solved the problem of the filaria's genetic cycle. First he showed that the embryos cannot develop in the same host that contains the adult forms. Among entozoons the eggs do not begin to develop until they leave the host inhabited by the parents. Similarly, the filaria embryos must escape.

This conjecture was based on an analogy, but it was transformed into a certainty by reasoning worthy of Harvey. Harvey had calculated that in one hour the left ventricle pumps an amount of blood through the aorta equivalent to three times the weight of the body. Where could such a large quantity of fluid come from, and where could it go? From this reasoning came the idea of circulation. Manson in turn calculated that at any given moment there were perhaps two million *Filaria immitis* embryos in the vessels of a canine host. What would happen if that number of animalcules began to grow while still in the dog's body? Before they had attained even a hundredth the size of the adult *Filaria*, their aggregate weight would exceed that of the dog. Such an absurdity would yield an even more absurd result: "The death of the host would imply the death of the parasite before a second generation of *filariae* could be born, and this of course entails the extermination of the species; for in such an arrangement reproduction would be equivalent to death of both parent and offspring, an anomaly impossible in nature."[13] In order to continue their development and ensure the propagation of the species, the embryos therefore must leave their first host. This leads to the idea of a cycle whose first phase is characterized by the exit of the microfilariae.

How does this exit take place? Generally, the eggs or embryos of the endoparasites are ejected in the process of excretion. They continue their development either in the environment into which they are deposited (*Ascaris lumbricoides*) or in the body of an animal that has fed on that environment (*Taenia*). Manson believed that the filaria embryos could exit by the same route, because several observers had reported finding the animalcules in the urine of chyluria victims. Ultimately, however, he rejected this mode of elimination because it depended too much on a morbid phenomenon, and the presence of the embryos was not always associated with manifestations of the disease. For Manson, an operation associated with the reproductive function could not be left to the mercy of an exceptional pathological phenomenon: "Nature is not likely to trust to the accident of a disease for the continuation of a species. Her operations are always orderly and reveal a plan. She may be careless of the single life, but she is very careful of the species." Because he believed in a wise and provident nature, Manson set about searching for an exit different from the route used by most endoparasites.

Manson also had another reason for bringing in an external factor. The embryos were not equipped to leave the circulatory system; they lived as prisoners in their sheaths, passively carried along by the blood. There were two possibilities: either the microfilaria were removed wholesale or they were removed piecemeal. If the former, they would have to be removed with part of the host's flesh. *Filaria sanguinis hominis* would then be like *Trichina spiralis*: "But in few countries is human flesh devoured in so wholesale a fashion as would warrant us supposing that *Filaria sanguinis hominis* was similarly treated." Manson rejected the analogy that rested on an assumption of cannibalism. Therefore the embryos had to be removed piecemeal.

Furthermore, the fact that the embryos reside in the blood was a clue. The first phase of the parasite's life cycle unfolds in this fluid medium. Hence the next phase must take place in the body of an animal that extracts them from that fluid—a bloodsucking animal. What parasites were always free and thirsty for blood? "Thus, then, the privilege will be confined to a very limited number of animals—the blood suckers. This includes the fleas, lice, bugs, leeches, mosquitoes, and sandflies."[14] The species to look at was the one that was prevalent where filariasis was found. The disease was confined to tropical and subtropical regions. The insect that removed the parasite from the host must exhibit a similar geographical distribution: it had to share the same habitat. Fleas, lice, bugs, leeches, and sandflies could be ruled out because they were found everywhere. That left the mosquito, or, more precisely, those species of mosquito that inhabited hot, humid regions.

In fact, two species of mosquito were frequently found in areas where filariasis was common. Manson chose the most common species, noting that the distribution of elephantiasis coincided with that of the *Culex* mosquito. Where the disease was prevalent, the insect abounded. In 1879 Manson was to prove that the microfilariae were adapted to the mosquito's nocturnal feeding habits. The embryos invaded the peripheral circulation at night, disappeared during the day, and then reappeared the following night. From this observation came Manson's famous "law of periodicity." It convinced Manson that he was right to have chosen the *Culex* as the filaria's intermediate host.

Manson already knew in 1877 that he had selected the right species. He allowed the insect to bite individuals whose blood was infested with embryos. In the mosquito's body he found four times as many larvae as in the same quantity of blood extracted with a needle: "From this it would appear that the mosquito has the faculty of selecting the embryon filariae; and in this strange circumstance we have an additional reason for concluding that this insect is the natural nurse of the parasite."[15] To trace the evolution of the microfilariae it remained to dissect mosquitoes at different intervals. Initially the filaria retains the appearance and motions it had while residing in its human host. Its sheath becomes more visible, and an oral rim appears. An hour after ingestion the filaria sheds its sheath. The mouth crease becomes more prominent, and the whiplike movements continue. Then the larval stage begins. Manson observed the separation of the tail and the first outline of a mouth. The body shrinks, thickens, and becomes transparent. Soon the alimentary canal appears. The final stage is characterized by accelerated growth. At the end of the sixth day the body attains its maximal thickness. At the end of a week the embryo ingested by the mosquito reaches maturity: "This formidable looking animal is indoubtedly the *Filaria sanguinis hominis* equipped for independent life and ready to quit its nurse the mosquito."[16]

At this point, however, a gap appears in the parasite's life history. How do the adult forms reach the lymphatic system of their final host? At the time it was believed that after the mosquito had ingested its meal of blood, it withdrew to a place near a body of water to digest its meal, deposit its eggs, and die: "Most mosquitoes die about the fourth or fifth day after feeding. . . . Perhaps death may not occur, as it usually does, soon after the eggs have been laid, and the insect may survive this operation for two or three days."[17] In any event, a few fortunate filariae attain the adult stage just as the mosquito dies. These filariae then devour the insect tissue and begin an independent life in the water, which presumably has a fortifying effect on them. The organism then somehow finds its way to the final host, probably by way of the drinking water: "The *Filaria* . . . escaping into the water in which the mosquito died . . . [and] being swallowed, it works its way through the alimentary canal to its final resting place."[18]

This conjecture was supported by an analogy as well as by consideration of certain cases of adaptation. In 1872 Fedschenko had shown that the Guinea worm completed its life cycle in the body of a tiny crustacean of the genus *Cyclops*, which was then swallowed along with the drinking water. The mosquito with its parasite cargo certainly was not swallowed; nevertheless, the filariae might well be ingested, as the crustacean was, along with the drinking water. So much for the analogy. As for adaptation, just as some instinct guided *Trichina spiralis* toward the muscles and *Filaria sanguinolenta* toward the esophagus, so, too, might some instinct guide *Filaria bancrofti* toward the lymph vessels, where they would mate, thus completing the cycle.

On June 20, 1879, the president of the National Board of Health approved the following resolution:

The memorandum of instruction for the commission appointed by the National Board of Health to visit the island of Cuba for the purpose of investigating certain points connected with the prevalence of yellow fever in that island states that in organizing the commission the National Board of Health has more especially in view the following desiderata:

First. To ascertain the actual sanitary condition of the principal ports in Cuba . . . and more especially as to what can and should be done to prevent the introduction of the cause of yellow fever into the shipping of these ports.

Second. To increase existing knowledge as to the pathology of yellow fever, that is, as to the changes and results which it produces in the human body.

Third. To obtain as much information as possible with regard to the so-called endemicity of yellow fever in Cuba, and the conditions which may be supposed to determine such endemicity. . . .

The three points above referred are believed to be those which will most certainly yield results of scientific investigation, and which therefore should receive the special attention of the commission.

But in addition to these the National Board of Health desires that the commission shall consider certain problems relating to this disease. . . . These problems relate to the nature and natural history of the cause of yellow fever, and the most important preliminary investigation on this point is to ascertain some means of recognizing the presence of the immediate cause of yellow fever other than the production of the disease in man.[19]

There were ample grounds for appointing such a commission. The previous year yellow fever had visited a hundred American cities. The effects had been disastrous: more than 120,000 victims, more than 20,000 deaths, and a monetary cost in excess of $100 million.

The commission left New York on July 3, 1879, and arrived in Cuba on July 7. It was to return on October 9 and submit a preliminary report on November 18.

The primary objective was to come up with an effective method of protection. In order to achieve that goal, the etiological agent had to be discovered. Given a disease whose cause was unknown, the commission adopted two working hypotheses. If the germ was a parasite, an effort should be made to find it both in the victim's body and in nature. The blood and affected organs of the victim were to be examined. If that examination failed to discover the parasite, an effort would be made to culture the germ from substances likely to contain it. If that, too, failed, then an attempt would be made at least to establish the germ's presence in the victim using a technique of communication: the injection of blood into animals. As for nature, both the water and the air should be examined. If the results were negative, further efforts would be made to establish the germ's existence in one or the other medium. The guiding principle of the investigation was that certain facts could be interpreted as if nature herself were spontaneously performing experiments.

The conclusions proved negative. In the end the commission would recommend employment of the usual preventive measures: anchoring ships offshore and confining crews to their vessels.

Consider first the research that was done on the victims of the disease. Most doctors viewed yellow fever as a disease of the blood. Jones claimed to have observed bacteria and filaments to which he ascribed a pathogenic role. And Richardson had reported a dumbbell-shaped bacteroid that he called *bacteria sanguinis febri flavo*. Examination of the blood thus became a confirming test. Such procedures were common at the time; Sternberg made them his specialty. Did he not confirm, the following year, the discovery of *Bacillus malariae*? Sternberg and Woodward used a microscope equipped with the finest lenses. Ninety-eight blood specimens taken from forty-one yellow fever victims were subjected to microscopic examination and photographed. The study revealed various remnants of epithelial cells, cottonlike threads, bacteria, and fungi. The presence of these common microorganisms showed that the precautions taken when the blood was drawn were not sufficient to prevent contamination by atmospheric germs. "We can claim no discoveries from the microscopic examination of the blood bearing upon the etiology of yellow fever."[20]

Did the pathogenic agent reside in organs exhibiting characteristic lesions? Satterhwaite and Richardson claimed to have observed spherical bacteria in the kidneys and liver. Guiteras focused his attention on those organs. The results were negative: "No microphytes have been found in the liver or in the blood contained in its vessels. . . . No organisms have been found in the kidney."[21] Oddly enough, the blood did not appear to be changed in any noteworthy manner. Examination of the blood during the course of the disease or post mortem showed no alteration of the corpuscles.

If yellow fever germs could not be observed directly, could they at least be cultured? Attention was focused on the blood, vomit, and urine of victims. Samples of each were cultured in coconut milk at room temperature. Examination of blood samples cultured in this way yielded nothing of note: only fungi and bacteria, the same microorganisms found in places where the disease had not broken out. Vomit and urine samples treated in the same way produced nothing unusual.

It remained to try to demonstrate the presence of the pathogenic agent in the blood. This was to be done by testing the susceptibility of various animals to the poison's influence. If an unambiguous case of the disease could be produced, it would at least prove that the poison was to be found in the circulatory system. Blood taken from victims was therefore injected into the femoral artery of a dog: "*Result*, entirely negative." A second experiment, identical to the first, also yielded nothing: "No bacteria were seen and no appreciable symptoms resulted from the injections."[22]

The commission therefore quickly turned its attention to study of the environment. Stagnant water had been suspected. The water in the port was consequently examined. Samples taken near the docks contained a considerable quantity of mineral matter, plant debris, infusoria, and bacteria. The microorganisms probably came from the sewers, but their reproductive powers and vitality were evidently impaired by the ocean salt: "That the water of the harbor generally is contaminated with this poison seems doubtful." The atmosphere had also been suspected. A series of experiments modeled on the work of Miquel, Tyndall, and Pasteur was therefore undertaken. Various methods were used to collect material suspended in the atmosphere: "The dust deposited was found to contain . . . in addition to the amorphous mineral and vegetable matters, epithelial cells, etc., which make up the greater part of the dust found everywhere."[23]

The Chaillé Commission was receptive to Finlay's old chemical theory. In 1858 Finlay had reported that the intensity of yellow fever seemed to coincide with the degree of alkalinity of the atmosphere.[24] Based on this single observation Finlay had constructed a medical theory. He envisioned a structural analogy between two mechanisms: the level of ammonia in the atmosphere was high because the poison was not being recycled, and similarly the level of ammonia in the human body was high because the poison was not being eliminated. This theory excited keen interest because it was interpreted in parasitological terms. The excess ammonia that produced the poisoning had to be a byproduct of the parasites' activities. Yet those parasites continued to elude detection.

It had yet to be shown that the germ was present in the atmosphere. Communicability of the disease was to be the key. "Which vessels," Chaillé asked, "are more apt to be infected, those at wharves or those at anchor?. . . Those at anchor near the shore or those more distant?" The answer was given by Burguess, the port's health inspector, who had collected various observational data that he was now able to interpret as an unplanned experiment: "All of my observations, as well as the facts recorded in the accompanying table, sustain me in the assertion that the liability to infection in this harbor is in an inverse ratio to the distance at which a vessel lies from wharves and habitations."[25] In short, the risk of infection increased with proximity to the wharves and diminished with distance.

At this point we are in a position to see clearly what facts were uppermost in Finlay's mind. The Plymouth Report raised the problem of the yellow fever germ's transformation outside the body. Manson had proposed a theory in which the mosquito served as intermediate host to microfilariae; the insect served as a first medium, enabling the embryos to exit their original host and reach a second medium, water, by means of which they made their way to their ultimate host. Finally, the Chaillé Report reported certain negative findings concerning the yellow fever germ. These elements, taken from Finlay's reading, guided him as he worked to unravel the mystery.

We can now retrace the crucial sequence of ideas. The first line of research concerned the transformation of the yellow fever germ outside the body, a subject broached by the Plymouth Report. Finlay would propose as solution of this problem the theory that Manson had just elaborated to explain the life cycle of the filaria. He would then refute this solution based on the conclusion contained in the

Chaillé Report. The second line of research concerned the problem of yellow fever's mode of propagation. As solution to this problem Finlay would also propose the theory that Manson had evolved to explain the transmission of filariasis. But Finlay again rejected this solution on the grounds that yellow fever, unlike filariasis, is not transmitted through the medium of water. Hence Finlay would abandon both these lines of reasearch. But not before reaping a certain theoretical profit: now he had in hand an object, the mosquito, of which it remained to form the concept. Perhaps this twin failure was a necessary condition for evolving the hypothesis that the *Culex* mosquito is the agent of transmission.

My objective is clear: to describe how Finlay found a way out of this double impasse. Knowledge is often the result of fruitless effort. To some this may seem surprising, but in the next chapter we shall dispel the apparent paradox and destroy a legend.

The Formation of a Hypothesis

The legend that must be laid to rest is this: that Finlay could not have made use of Manson's work because he was unfamiliar with it. If true, this would of course undermine my historical reconstruction. I shall therefore prove that Finlay was familiar with Manson's work. Manson's work therefore must have been known, and disseminated, well before 1881. I shall also show that Finlay's silence was a matter of dissimulation. And finally I shall attempt to clarify the reason why Finlay thought it wise not to acknowledge his debt.

A set of problems that Finlay deliberately left in obscurity must be brought to light. Of course a scientist is under no obligation to explain the gropings, reverses, and dead ends that mark the course of research. But where Finlay's obligation ends, the work of the historian begins. It is the historian's task to reveal how he formed his hypothesis, or, to put it another way, to describe the mind's progressive embrace of the mechanisms of nature.

The hypothesis grew out of Finlay's reflection on the problem of how the yellow fever germ is transformed outside the body. It was crucial to ask where the modification of the germ might take place, for only then was there reason to look for a novel answer—the mosquito as intermediary host—in a neighboring field of inquiry. Finlay quickly rejected Manson's solution, but the episode nevertheless focused his attention on the Amoy physician's theory.

The second problem concerns the mode of propagation of yellow fever. This time, the attempt to resolve the problem by applying Manson's theory led Finlay to center his interest on one of its elements: the mosquito.

The insect thus became caught up in a set of questions involving parasites with alternating hosts. Finlay, however, incorporated the mosquito into an epidemiological theory, thereby giving it the status of a medium (as opposed to a host). Finlay thus abandoned the path followed by Manson and mapped out one of his own. This led him to initiate the research that led to his hypothesis that the *Culex* is the agent of transmission in yellow fever.

AN IMPLICIT PROBLEMATIC

Finlay formed his hypothesis at the end of 1880. Before that time one searches his work in vain for the slightest allusion to the mosquito. In the summer of 1879 Finlay presented a refurbished version of his old alkalinity theory. Two subsequent papers, written before the summer of 1880, limit themselves to defining the measures to be taken in order to conduct a rational study of the disease. January 1881: en route to an international conference in Washington, Finlay proved that the *Culex* mosquito may consume more than one meal of blood. A female, confined after fertilization, bit twelve times and laid eggs three times before dying in the United States, where the temperature fell below freezing. This observation lent credence to the notion that the mosquito could be the agent of transmission. It satisfied the required condition, namely, the ability to bite *more than once.*

In February 1881 Finlay communicated his hypothesis, but in a form so enigmatic that it would not be understood: he said simply that for the disease to be transmitted from a victim to a receptive individual one had to assume the intervention "of an agent entirely independent for its existence both of the disease and of the sick man."[1] It would be churlish to blame the participants in the Washington conference for their failure to understand the Cuban physician's hypothesis. Finlay did not say what this agent was; he did not name the mosquito. Furthermore, as Da Silva Amado points out, "the conference, composed largely of diplomats, showed a certain disdain for scientific questions."[2] Finlay nevertheless had elaborated his hypothesis some time between the summer of 1880 and January 1881. By his own admission, it was "in December 1880" that he was led "to believe that the sole mode of transmission . . . had to be inoculation . . . through the intermediary of some biting insect."[3]

Historians of medicine have frequently raised the question of how Finlay conceived the hypothesis of the mosquito as agent of transmission. Here is the answer. Finlay found in the Plymouth Report the theory of the nidus: germs stemming from the disease victim undergo transformation in the atmosphere, which also serves as a vehicle. Against this theory Finlay noted Burguess's observations in the Chaillé Report: the atmosphere could not have been the germs' vehicle, since vessels not in communication with the shore were not infected. Hence the atmosphere was not the nidus. Concerning the question of the germs' transformation outside the body, Finlay gave a new answer: Manson's theory of the life cycle of filaria. He substituted the mosquito for the nidus. But he then immediately rejected that answer as well owing to the negative conclusions contained in the Chaillé Report: the yellow fever germ could not be located. It would have been presumptuous to assert that the mosquito was the intermediate host of an unknown parasite.

Finlay then turned his attention to the only problem

susceptible of analysis, namely, the propagation of the disease. To solve it, Finlay would once again apply the theory Manson had proposed to account for the transmission of filariasis: the mosquito, laden with its cargo of filariae, dies in water; the filariae then escape and are immediately swallowed by man along with the contaminated water. Finlay, however, was in a position to show that, unlike filariasis, yellow fever did not have water as its ultimate medium. The mosquito was therefore useless. Its role could be restored, however, if it was assumed that after feasting on infected blood it went on to bite a receptive individual. From this reasoning emerged the hypothesis of the *Culex* mosquito as the agent of transmission.

Cuban historians will find this reconstruction hard to accept. That is not surprising. They have worked hard to establish the legend of Manson as a scientist unknown to the European scientific community and to Finlay. Manson worked in China. All the accounts have to one degree or another embroidered on this central fact. In 1935, for example, Dominguez Roldan observed: "In 1879 Dr. Patrick Manson discussed the indirect transmission of filaria by way of the mosquito, but . . . an account of Manson's work did not reach scientific centers of Europe and America until much later; hence neither [Manson nor Finlay] was able to influence the work of the other."[4] In 1954 Hurtado Galtes insisted: "These two inspired researchers were contemporaries, but since they worked at opposite ends of the earth, neither was able to influence the other."[5] More recently Delgado Garcia has written: "It was after Finlay described his verified theory in 1881 that Manson, in 1883, finally completed the life cycle with the mosquito *Culex fatigans*."[6] The aplomb of these Cuban historians is remarkable; their unsubstantiated assertions are obviously intended to rule out the notion that Manson influenced Finlay, thereby ensuring the priority of the latter.

These assertions are historically false. Manson's work

soon became known to the European medical community. Distance was no obstacle to the diffusion of Manson's results in England and Cuba. It is true that Manson worked in China and that initially he was alone. But it is no less true that he soon found a way to break out of his solitude. Cobbold was in a position to know: "No doubt it was a feeling of isolation that at length induced Dr. Manson to make me the instrument of bringing his later researches before the public."[7] According to Alcock, Cobbold acted as Manson's "active publicity agent." Accounts of Manson's work and excerpts from his reports were published in scientific journals. Diffusion was rapid. Consider one example. On January 4, 1878, Cobbold received from Amoy a letter dated November 27, 1877. In the letter Manson announced his discovery of the intermediate host of *Filaria sanguinis hominis*. Cobbold published the news in the weekly *Lancet* dated January 12, 1878. That issue reached Cuba in February. Thus a discovery made in Amoy could have been known in Havana eight to ten weeks later. *The Lancet* was in fact one of the primary sources of information available to Cuban physicians. It was particularly useful to Finlay, who of course read English fluently.

Finlay made use of Manson's work in December 1880, hence he must have become aware of it before that date. When was Finlay's interest directed toward the Amoy physician's work? The contagion of leprosy was discussed at Havana's Academy of Sciences on two occasions. In November 1879 Finlay's contribution to the discussion showed that he believed, but not without misgivings, that leprosy is a contagious disease: "If the disease is in fact contagious, a tropical climate must also be necessary in order for contagion to occur."[8] In January 1880 Finlay again spoke out in favor of contagion. This time, however, he insisted that "a number of necessary external conditions" must be met.[9] Finlay's first statement is to his second as a superficial perception is to a mature appreciation of the difficulties involved in understanding indirectly transmissible diseases.

One possible reason for this change in perspective was that Finlay in the interim had researched leprosy in the medical journals, including *The Lancet*. Tracing the subjects of leprosy and elephantiasis Graecorum would have led him to filariasis. In an article by Fayrer that appeared in February 1879 Finlay could have read the following: "It has been shown that disorders of the lymphatic system are most frequently associated with, if not caused by, the filaria; also naevoid and ordinary elephantiasis Arabum; perhaps also elephantiasis Graecorum." This passage refers the reader to a note stating that Joseph Bancroft had found filariae in the blood of a leper.[10] A few pages later Fayrer notes that the transmission of filariasis requires a "combination of different conditions."[11] In January 1880 Finlay no doubt applied to leprosy what he had just learned about filariasis. Naturally I would not have ventured this hypothesis in the absence of proof that Finlay was familiar with Manson's theory. That proof may be found in the 1881 memoir, where Finlay in effect makes reference to the Amoy physician's theory. But he does so in a context likely to render that reference unrecognizable, so it is not surprising that it has gone unnoticed. I shall come back to this point in a moment.

There is more. One year after formulating his hypothesis, four months after reading his 1881 memoir to the Washington conference, Finlay embarked on research on filariasis. In March 1882 he stated that he had observed several cases of the disease the previous January and that he had confirmed Manson's law of periodicity. This paper, too, shows that Finlay had indeed read and drawn upon Fayrer's long article in the February 1879 *Lancet*, entitled "On the Relation of *Filaria Sanguinis Hominis* to the Endemic Diasease of India." This accounts for the references to Manson and for the following remark: "We are indebted to him for an ingenious theory, according to which the transit of the larvae through the mosquito's body after the mosquito has sucked the blood in which they were living is considered to be a necessary phase in the filaria's evolution."[12] In conclusion, Finlay recommends that his colleagues examine the texts for themselves: "Those interested in further information on the question should go directly to the sources and consult the interesting articles that have been published in *The Lancet* of London over the past ten years."[13] Insofar as Finlay's research is based on Manson's, Finlay's interest in filariasis in 1882 is certainly no accident. Let me sum up what has been shown thus far. Manson's work was divulged publicly between 1878 and 1880, and distance was no obstacle to its being reported in Havana. Finlay had access to information about that work. He probably became aware of it during the discussion of leprosy's contagiousness, at which time he still had no idea that he would later, in December 1880, make use of it in his own work.

Manson's work became relevant to Finlay's in December 1880 because that was the point at which Finlay needed to confront the question of the transformation of the yellow fever germ outside the body and come up with an alternative to the nidus theory. The usefulness of Manson's theory became apparent on the day Finlay posed that question, raised in his mind by the Plymouth Report. That report was of course submitted to the United States Navy Department in May 1880 and published the same year. Finlay probably read it in December. What is more, he several times stated how greatly the report had interested him. More than that, he indicated that what had most attracted his attention in the report was the question of how the yellow fever germ is transformed outside the body. Consider, for example, his recollection of how he discovered the mosquito in 1880: "This happened about the time when Bemiss, Stone, and other American yellow-fever experts had invented the 'nidus theory.' . . . I had however conceived a different solution of the problem. My own conclusion had been that . . . the natural transmitter of the disease must be a blood-sucking insect."[14]

Finlay, however, never explained how he went from the question of transformation outside the body to the mosquito theory. This gap in his reasoning is not without guile. We must reinstate the steps that Finlay deliberately left out. And since those steps depend on Manson's theory, we must first explain why the Cuban physician carefully concealed his debt.

Part of the answer is to be found in the way in which Finlay judged Manson's work after 1882. Finlay cited an observation that he believed contradicted Manson's theory: "I have grounds for not accepting Dr. Manson's theory. I base this conclusion on, among other things, the fact that filariae in the blood of one of my patients had attained a stage of development that, according to the theory in question, they should have attained only in the stomach of the mosquito." In addition, Finlay cited Fedschenko, who had discovered the life cycle of the Guinea worm, as a forerunner of Manson. This was the reason for his seemingly indignant reaction: "I cannot, however, help protesting against the foolishness of certain authors who, without conducting experiments of their own, characterize Dr. Manson's theory as a fiction."[15]

How could Finlay challenge the validity of Manson's theory by adducing a contradictory observation while *at the same time* defending its authenticity, presumably acknowledging that it possessed some value? Expressing doubts about the validity of a theory may mean that one welcomes the theory but with certain reservations. By stating that he does not accept Manson's theory, Finlay dissuades us from thinking that he might owe Manson anything. He suggests that is would be a mistake to look to Manson's work for any kind of antecedent to his own. But

THE FORMATION OF A HYPOTHESIS

to portray Fedschenko as Manson's forerunner is to diminish Manson's merit. Finlay unwittingly allows us to see how he situates his own work relative to Manson's. Since Manson preceded him in pointing a finger of suspicion at the mosquito, the *Culex* hypothesis was not as original as Finlay would have us believe.

The reason for Finlay's conduct is clear. By eliminating Manson's influence, he conjures away his own greatest fear: that he will be judged a figure of secondary importance. This explains his choice of tactic: the less he speaks about the Amoy physician, the more novel his own work will seem. In any case, Finlay's ambivalence toward Manson is a clue: it focuses attention on a *determinative relation*—one that Finlay could not avow.

It is now clear why Finlay did not refer to Manson's work. Finlay's dissimulation gave rise to an insurmountable difficulty, however. By concealing his borrowing, he made it impossible to explain the progress of his own thinking. Historians who have not noticed his tactic have not been able to do much better than Finlay himself. They have all run up against the same obstacle. The result has been any number of histories condemned to platitude because they were content to repeat explanations so dubious that I am still astonished they were not more skeptically received.

Now that we are liberated, I shall move without further ado to the consequences. Finlay's texts, as is only to be expected, are full of gaps and weak arguments. The gaps must be filled in and the deceptions uncovered. We must point out where Finlay substituted inconsistent reasoning for a clinching argument or pertinent reference. The danger, of course, is that in attempting to retrace Finlay's line of thought we may end up with an inaccurate reconstruction. To avoid this danger, I shall support my case with reference to Finlay's texts. A further difficulty arises. Finlay, as we just mentioned, left gaps in his arguments and concealed his borrowings. But the difficulty is more apparent than real. The 1881 memoir and many subsequent articles contain enough information to cast a revealing light on the progress of his thinking. I shall examine his texts in the manner of an investigator. By comparing testimony, contrasting statements, and examining clues I shall seek to bring to light what Finlay sought to conceal. With care and patience the mystery can be unraveled.

THE INTERMEDIATE HOST

Finlay's first line of research concerned the transformation of the yellow fever germ outside the body. This problem does not figure in the 1881 memoir, however. Finlay left it out, as he was well aware: "By reason of other considerations which need not be stated here, I came to think that the mosquito might be the transmitter of yellow fever."[16] There are gaps in the 1881 memoir, and there is a certain consistency in what is left out. By not mentioning the extrabodily transformation, Finlay relieved himself of the obligation to discuss Manson's theory and his reasons for rejecting it. It is true that Finlay would soon come to regard this question and its answer as irrelevant to the matter at hand. Nevertheless, they were essential to the formulation of his hypothesis. In 1901, in a paper presented to the Pan-American Medical Congress, Finlay alluded to what he had omitted to mention in 1881. I shall use this text to elucidate his initial line of research.

As mentioned earlier, most American physicians believed that the pathogenic agent initially resided in the environment. Bemiss was one of the few who believed that it emanated from the patient. In any case, everyone agreed that heat, humidity, and unsanitary conditions contributed to the effect of the germs. Neverthless, as Finlay was well aware, even when these favorable circumstances were found, the disease sometimes failed to break out: "It was therefore necessary to search for an additional factor that might account for this peculiarity. In my opinion, the first attempt to get around this problem was contained in a note of the report on the U.S.S. Plymouth, citing the opinion of Dr. Bemiss that the yellow fever virus . . . after leaving the body must undergo certain changes to increase its toxic properties." In assuming the existence of an environment apt to encourage the transformation of the germs, Bemiss focused attention on a range of factors, some known, others yet to be discovered. Bemiss also pointed to the atmosphere as the germs' vehicle; he spoke of inhalation of the poison via the respiratory system.

In Finlay's eyes, this theory left two problems unresolved. First, it said nothing specific about the nature of the nidus. "This ingenious theory failed, however, to explain what conditions might favor or inhibit the supposed transformation of the humanized and inert germ." Second, it had nothing convincing to say about the mode of transmission: "The theory also avoided previous difficulties: How was the germ introduced into the body of the healthy subject?"

Where the American doctor believed he was offering acceptable solutions, Finlay saw equivocation. The "testimony of Dr. Burguess" indicated some of the weaknesses: "It was believed that the only possible means of contamination was inspiration via the respiratory system. But this mode of propagation seemed unsatisfactory to me, in that it assumed diffusion of the active germ through the atmosphere, which contradicted the observed fact that the winds had little or no influence on the propagation of the epidemics." Disproof of the atmospheric mode of transmission amounted to refutation of the nidus theory. If the atmosphere was not the germs' vehicle, they did not reside in the air. The nidus theory did not just suffer from short-comings and vagueness; it was simply wrong. Not only was the atmosphere not the vehicle, it was not even the environment in which the transformation of the germs outside the body took place.

"Based on these considerations, I came to the conclusion that the yellow fever germ must be introduced into nonimmune victims by way of inoculation. . . . I thought of the mosquito." A clarification follows: "Initially, I did not think that the supposed transformation of the human yellow fever germ outside the human body (or perhaps inside the mosquito's body) was necessary."[17] These remarks concerned the refutation of the nidus theory. The clarification assumes the refutation of the notion that the mosquito is the intermediate host of the yellow fever germ. Only a solution deemed irrelevant can be rejected, but in order to know that it is irrelevant it must have been tested. If we can now demonstrate the link between the remarks on the nidus theory and the clarification, Finlay's initial line of research will become clear. After rejecting the nidus theory, Finlay used Manson's theory to solve the problem of the germ's transformation. Then he rejected it, too.

Finlay surely made a crucial choice in focusing his attention on the Bemiss version. Those who believed that the germs reside initially in the environment were forced to look upon yellow fever as an *infectious* disease and to conceive of the nidus as an inanimate environment: investigation was then limited to physical and chemical factors. By contrast, if one assumed that the germs emanated from the disease victim, yellow fever then belonged to the class of *indirectly tramsmissible* diseases. The researcher was therefore confronted with this alternative: whether to envision the environment as animate or inanimate. Bemiss opted for the inanimate environment because he assumed that the yellow fever germ was a microphyte. He therefore looked for factors likely to influence the evolution of plants—once again, physical and chemical in nature. By opting for an animate environment, Finlay administered a fillip to classical epidemiology. Nevertheless, he was forced to abandon the notion that the germ emanates from the victim and to look to research in a peripheral field for another solution. Such a solution was provided by Manson's work on the filaria.

Finlay first asked under what conditions it would be legitimate to apply this solution. What if the fever germ were not a nematode but an identifiable microorganism in the blood? Unfortunately, microscopic examination of the blood revealed no such germ. Finlay thus responded to the negative conclusions of the Chaillé Report. The fact that the yellow fever germ was unknown was sufficient reason to reject Manson's solution. The problem of the host could not be formulated, let alone resolved, until the question of the parasite was answered. At the time, Finlay could not accept the idea that the mosquito served as host to an unknown germ; he also came to regard the question of transformation of the germ outside the body as premature.

Finlay's abandonment of this initial line of research had a positive aspect. Pursuing that line had focused his attention on Manson's theory. Applying that theory to the problem of yellow fever's propagation brought Finlay one step closer to his discovery.

TWO MEDIA

Finlay was not very explicit about his second line of research. From our standpoint, however, the more enigmatic his manner of expression, the more clearly he reveals himself. The paradox is illusory. Finlay was clear because he was no longer able to hide the fact that he had used Manson's theory. So much is evident once we recognize that Finlay tried to make his reasoning seem perfectly natural. Consider this passage:

Let us consider by what means that insect might transmit the yellow fever, if that disease happens [not] to be really transmissible through the inoculation of blood. The first and most natural idea would be that the transmission might be effected through the virulent blood which the mosquito has sucked, amounting to 5 and even to 7 or 9 cubic millimeters, and which, if the insect happens to die before completing its digestion, would be in excellent condition to retain during a long time its infecting properties. It might also be supposed that the same blood which the mosquito discharges, as excrement, after having bitten a yellow fever patient might be dissolved in the drinking water, whereby the infection might be conveyed if the latter were susceptible of penetrating by the mouth. But the experiments of Ffirth and other considerations arising from my personal ideas regarding the pathogenesis of yellow fever forbid my taking into account either of those modes of propagation.[18]

I see difficulties of two kinds in this passage: syntactic and semantic (or historical). First consider the syntactic difficulties, because they alter, or, rather, obscure, the meaning. Finlay's conditional is rather puzzling: "Let us consider by what means that insect might transmit the yellow fever, if that disease happens to be really transmissible through the inoculation of blood." He then goes on to consider two possible means of transmission via the drinking water, that is, *not* involving inoculation. It would make more sense if the sentence in question read " if that disease happens [not] to be really transmissible through the inoculation of blood."

This defective sentence was, if not pointed out, nevertheless noticed by Matas, who in 1882 translated Finlay's memoir. Matas was of course familiar with both Spanish and English; he had been the Chaillé Commission's official translator. His version reads as follows: "We ask ourselves, supposing yellow fever to be an affection contagious through the blood, by what means could the mosquito transmit the disease."[19] On the one hand Matas eliminates the defective construction. On the other, he interprets Finlay's assertion in a way that is not entirely faithful to Finlay's meaning. The generality "contagious through the blood" says nothing about the means of transmission, "inoculation." It is preferable to retain Finlay's wording but to restore the negative: "If that disease happens [not] to be really transmissible through the inoculation of blood."

As for the semantic difficulties, they are fairly easily resolved. Finlay's first possibility reproduces Manson's theory: the blood-laden mosquito dies in the water before digesting the blood. Finlay gives a variant of Manson's theory when he describes the second means of transmission: the mosquito digests the virulent blood and then discharges it as excrements into the water, which the victim then swallows. This second possibility leaves no doubt about the implicit mechanism of the first means of transmission: the mosquito dies in the water, which is then contaminated by the virulent blood it has taken from a victim of the disease; the contaminated water is the ultimate medium of transmission.

Some historians may still find it difficult to accept that it is indeed Manson's theory that appears in the 1881 memoir. The proof lies in a comparison between what Finlay wrote and what he might have read in various articles in *The Lancet* published before 1880. I shall mention just a few examples. Fayrer (March 16, 1878, p. 376a): "Dr. Manson followed the progress of the embryo from the body of the mosquito to the water and from there back to man, reproducing the cycle of genetic change." Fayrer (February 8, 1879, p. 190a): "Concerning the mode of entry into the human body, Manson said, speaking of the embryo, that after it left the mosquito and escaped into the water, 'upon being swallowed it made its way down the alimentary canal.' " Fayrer (February 22, 1879, p. 269a): "At the end of the fifth day, the larvae escaped from the mosquitoes into the water and were then swallowed by man."

Finlay was enigmatic, however, because he mentioned Manson's theory in a context that suggested he had *already* elaborated his own hypothesis. Consider again the sentence: "Let us consider by what means that insect might transmit the yellow fever, if that disease happens [not] to be really transmissible through the inoculation of blood." On the one hand this proposition legitimates the question that will allow Finlay to introduce Manson's theory. On the other hand, by omitting the negative, it suggests that in Finlay's view inoculaton is in fact the means of transmission. The omission of the negative is therefore remarkably revealing, particularly since in the preceding pages Finlay has shown that the *Culex* mosquito satisfies all the conditions required of an inoculatory agent.

Manson's theory is therefore presented in the guise of a *rival* hypothesis. It is presented, moreover, as though it were a theory of Finlay himself, but one for which he deserves no credit, for what he says seems self-evident: either the mosquito dies quickly and contaminates the water with the victim's blood, or it survives and contaminates the water with excrement resulting from digestion of the same blood. Both means of transmission are considered solely in order to be dismissed, without profit. Finlay counters with his own hypothesis: "My personal ideas concerning the pathogenesis of yellow fever," which accord with the notion that the mosquito is the inoculatory agent. In effect, the lesion of the vascular endothelium is the *signature* of the aggressive agents of the disease. This leads to the idea that the virus is transported from the lesioned tissue of a victim to the corresponding tissue of a healthy individual by way of the insect. Finlay thus opposes Manson's theory with a theory of his own: yellow fever is transmissible by the bite of a mosquito.

It is now time to lay to rest once and for all the objection that Finlay could not have made use of Manson's work. This objection would carry weight if what Finlay wants us to believe were in fact true: first, that from the beginning he looked upon the mosquito as the agent of transmission; second, that he invented two rival hypotheses; and third, that he refuted both these hypotheses without benefitting from them in any way. The first assertion is incorrect, because Finlay's initial concern was with the transformation of the yellow fever germ outside the body. We saw earlier that Finlay, following Manson, first introduced the mosquito as an intermediate host. It follows that the second assertion must also be dismissed. Finlay did not reinvent Manson's theory; rather, he used it to resolve the problem of transmission. Furthermore, this theory, which combined two media (mosquito and water), arose from the study of filariasis. It is hard to see how Finlay could have come up with it other than through the study of worm-related diseases. Finlay may have refuted Manson's theory, but he profited from his awareness of it.

It is important to establish the order of factors. In reality, the theory borrowed from Manson constituted Finlay's first solution to the problem of yellow fever's transmission. He reaped the benefit of his critical examination of that theory. It was after analyzing Manson's solution that Finlay came to discover the mosquito. But let us not get ahead of ourselves.

What justified the application of Manson's theory to the problem of yellow fever's transmission? How did Finlay refute the solution derived from that application? The assumption that justified the use of Manson's theory actually pointed the way toward its refutation. Finlay rejected the mosquito as intermediate host for the justifiable reason that the yellow fever germ was unknown. We may then ask why Finlay persisted in using Manson's theory a second time? As long as the nature of the yellow fever germ remained unknown, it could not be maintained that the mosquito serves as intermediary host. Finlay not only avoided making any such assertion but actually rejected it. Simply because the yellow fever germ is unknown, however, is no reason to deny that it is present in the blood of victims. Finlay explicitly stated that it was present in the blood. There was nothing extravagant about this conjecture, moreover, because there existed another disease whose germ, though still unknown, was believed for good reason to reside in the blood: cholera.

If for the sake of argument one admitted that the yellow fever germ resides in the blood, then it was legitimate to think that yellow fever might be transmitted in a manner similar to filariasis. In other words, it was legitimate to think that yellow fever might be transmitted through contaminated water by means of virulent blood taken from a victim by a mosquito that then died or by means of excrement made from such blood if the mosquito continued to live. It was equally legitimate, again assuming that the yellow fever germ resides in the blood, to think that yellow fever might be transmitted in a manner analogous to cholera, that is, through water contaminated by the waste of disease victims.

Now things begin to be clear. If yellow fever was transmitted in the manner of filariasis, the medium of transmission was ultimately the same as for cholera, namely, water. Both the yellow fever victim and the mosquito behaved like the cholera victim. The yellow fever victim's waste might contaminate the water directly. The mosquito, moreover, might, if it died quickly, contaminate the water with blood taken from the victim, or, if it continued to live, it might contaminate the water with excrement stemming from that blood. Manson's theory applied to yellow fever replicated the transmission of cholera: the mosquito was merely a "dynamo" of dissemination. We can now see why Finlay supplemented Manson's theory (that the mosquito dies in the water and releases the filaria-containing blood) with a variant of his own devising. Finlay theorized that the yellow fever germ, like the cholera germ, is not altered by digestion, from which it followed that the mosquito could contaminate the water through its excrement.

Let us be careful not to lose sight of the essential point. If the virus is in the blood, the yellow fever victim and the mosquito that bites him inevitably behave like the victims of cholera. This assumption leads to the refutation of Manson's theory. Is yellow fever transmitted, like cholera, via a morbid principle contained in the blood or excrement? If not, it follows that yellow fever is not transmitted in the same way as filariasis. Now, it happens that certain aspects of yellow fever's epidemiology combined with certain experiments of Ffirth to yield a negative answer to this question.

Consider the epidemiological issues first. Parkes stated that the germs of yellow fever were to be found in the excrement of victims; like cholera, yellow fever was a "fecal disease." The Chaillé Commission, impressed by this statement, asked the Cuban physician to substantiate it. Finlay was no amateur. In 1868 he had discovered irrefutable evidence of a correlation between the incidence of cholera and the structure of the water distribution system. Finlay's work had been a remarkable application of the ideas of Snow. His epidemiological study of yellow fever, however, showed that there was no relation between the emptying of cesspools and the propagation of the disease: "From these facts it would seem that yellow fever is not a fecal disease."[20] Most American experts in any case agreed that yellow fever was not propagated in the same way as cholera. Finlay was aware, moreover, that those same experts had eliminated drinking water as a vehicle of dissemination. This widespread idea underlay Finlay's conviction that neither the yellow fever victim *nor the mosquito* resembles the cholera victim.

We come now to the experiments of Ffirth. These involved attempts to produce contamination by ingestion of waste matter or vomit taken from yellow fever victims or by injections of blood serum. It is tempting to compare these attempts with attempts to induce cholera artificially, attempts with which Finlay, having written a well-documented 1873 article on the subject, was thoroughly familiar. The experiments with cholera had yielded positive results: "Cholera is transmissible through a specific substance contained in intestinal evacuations, vomit, and blood serum."[21] By contrast, the experiments with yellow fever had proved negative: yellow fever was not transmissible through a morbid principle contained in excretions, secretions, or blood serum.[22] Finlay therefore believed that yellow fever was also not transmissible through water contaminated by the blood and excrement of victims or by blood and excrement from mosquitoes that had bitten victims. Whereas Ffirth concluded that yellow fever was not contagious, Finlay deduced that it was *not indirectly transmissible in the manner of cholera*.

It is now clear why Finlay in his 1881 memoir did not invoke epidemiological considerations in opposition to Manson's theory. To have done so would have revealed the logic behind his refutation and would therefore have disclosed how he came to his discovery of the mosquito. Indeed, to argue that yellow fever is not transmitted in the same way as filariasis (that is, through water contaminated by mosquito blood or excrement) *because it is not transmitted in the same way as cholera* (that is, through water contaminated by the blood and excrement of the victim) would have led inevitably to the question, What is the purpose of the mosquito?

Finlay therefore focused attention exclusively on the experiments of Ffirth. By invoking them, without further clarification, in criticism of Manson's theory, Finlay was able to reject the theory without providing the logic behind the refutation. He remained deliberately allusive and obscure. Allusive, because he did not discuss the content of Ffirth's experiments. Obscure, insofar as he failed to state that, to him, those experiments demonstrated that yellow fever is not transmitted in the same way as cholera.

Finlay was obliged to conceal the logic of refutation because he would have been forced otherwise to admit facts that he was trying to hide: that Manson's theory constituted his first solution to the problem of yellow fever's transmission, and that the refutation of that solution provided him with the key to the problem. Manson's theory constituted Finlay's first solution because, once it was assumed that the yellow fever germ is found in the blood, it was easy to draw an analogy between the transmission of filariasis and that of yellow fever. This assumption not only justified the importation on Manson's theory but also allowed Finlay to draw a further analogy between the transmission of yellow fever and that of cholera. Second, the refutation of the Manson-derived solution to the problem provided Finlay with the key to his own solution: once it had been shown that yellow fever is not transmitted in the same way as cholera, Finlay was able to put his finger on that part of Manson's theory that was also involved in the transmission of cholera, namely, the contaminated water. From this came the idea that the mosquito does not behave in the same way as a cholera victim.

In the overall economy of Finlay's analysis this refutation is simply the negative counterpart of a far more subtle positive component: Finlay retained the mosquito even after he had rejected its function. Historians who have failed to notice Finlay's critique of Manson have not surprisingly also failed to notice the positive element. The time has now come to see how Finlay endowed the mosquito with a new function: neither intermediate host nor primary medium but medium in the true sense of the word.

THE AGENT OF TRANSMISSION

Finlay separated the doublets intermediate host/mosquito and medium/water; having rejected the intermediate host and the water, he combined the two remaining possibilities and came up with the hypothesis that the mosquito is the medium. He did not choose the mosquito from among the various bloodsucking insects. Out of this conceptual work came a research program: to take a fresh look at the problem of the yellow fever germ's location, to embark on research in entomology, and to perform corroborative experiments.

I shall first show that Finlay did not choose the mosquito but rather took it from Manson. I shall then show how he reformulated the problem of the locus of the morbid principle because the localization of the virus in the blood did not square with the role Finlay attributed to the mosquito. From this came the new conjecture concerning the location of the virus, which was intimately related to the notion of the insect as medium. Finally, I shall describe how the mosquito's hypothetical role as agent of transmission made it necessary to choose a new object of study and to carry out a new line of experiments. A new object of study, because the hypothesis that the mosquito is the vehicle of the disease required that the mosquito bite more than once. This led Finlay to investigate the biology of the mosquito. And a new line of experiments, because the only way to demonstrate the insect's role was to produce cases of the disease experimentally.

That Finlay did not single out the mosquito from among the bloodsucking insects can be seen by comparing two texts. In the 1881 memoir Finlay wrote: "Once the need of an agent of transmission is admitted . . . it seemed unlikely, therefore, that this agent should be found among [microphytes],[23] for those lowest orders of animal life are but little affected by such meteoric variations. . . . I came to think that the mosquito might be the transmitter of yellow fever."[24] In 1886 Finlay wrote: "The history and etiology of yellow fever exclude from our consideration as possible agents of transmission other bloodsucking insects such as fleas, etc., the habits and geographical distribution of which in no wise agree with the cause of that disease."[25]

In 1881 Finlay gave the authentic version: he had chosen between microorganisms (microphytes, microzoa) and the mosquito. In 1886 he gave a refurbished version, a correct and convincing argument according to which the mosquito was chosen from among other bloodsucking insects such as fleas. The reason why Finlay made this belated correction is that he had spotted the weak point in the initial argument. The mention of microorganisms as a possible choice was not merely unfortunate but also revealing. The author of this strange hypothesis was of course the American, Jones: "Such poison may be conveyed by minute forms of plant or animal life . . . conveying the poison by being wafted about by currents of air."[26] in 1881, therefore, Finlay had been faced with a choice between Jone's theory and Manson's. The former was susceptible to the same criticism as the nidus theory: just as the atmosphere was not the vehicle of yellow fever germs, neither was it the vehicle of microorganisms mentioned by Jones. That left Manson's theory. From the first, therefore, Finlay had proposed the mosquito.

Having concealed his borrowing, Finlay was led to justify his choice of the mosquito by linking it to earlier observations. He therefore quoted excerpts from Brehm's *Les merveilles de la nature*, La Roche's *Yellow Fever* (1853), and Humboldt and Bonpland's *Travels* (1814).

From the first of these works he took observations of Aristotle, Pliny, and Pausanias concerning the proboscis of the mosquito. From the second he took a remark made by Rush concerning the unusually large number of mosquitoes in Philadelphia at the time of the 1797 epidemic. From the third he took remarks concerning the habits of mosquitoes and early historical accounts of the torments they inflicted on Cortez's men. Finlay thus posited a connection between historical and ethnographic accounts and a choice of object dictated by a specific problem. We must not allow ourselves to be misled by this alleged connection, for that would obscure the fact that Finlay is actually concealing his debt to Manson. It would then be possible to say, along with Blanchard, that "an extreme pullulation of mosquitoes was noted in several major epidemics, as in Philedelphia in 1797. Struck by this fact . . . Finlay had the idea that the disease was in fact inoculated by mosquitoes."[27]

Finlay was forced to reconsider the localization of the yellow fever germ and the circumstances of its transmission. The claim that the mosquito functioned as medium raised a difficulty. Finlay had eliminated the possibility that yellow fever is transmitted in a manner similar to filariasis by showing that the virus did not reside in the blood. In order to cast the mosquito, a bloodsucking insect, in the role of vehicle, a new conjecture concerning the seat of the virus was required. If the germs were not in the blood, they must lie somewhere along the path of the puncture. The virus was in fact removed along with the blood but from the vascular endothelium, where it lived and multiplied. In order for the disease to be transmitted, one condition had to be satisfied: the virulent matter had to be inoculated into the corresponding tissue of a healthy individual. "It is not from the blood that has been sucked that the mosquito is supposed to derive its contamination, but from the tissues that the sting must bore through. . . . The proboscis of the

Culex, laden with contagious material, then penetrates the skin of a healthy individual and deposits the morbid particles into the tissue that will serve as their medium."[28]

This hypothesis concerning the seat of the morbid principle was immediately incorporated into a corroborating pathogenic theory based on the assumption that yellow fever was one of the zymotic fevers, an eruptive fever whose eruption was located in the vascular endothelium. Pathological anatomy, by revealing a fatty degeneration of the capillary walls, seemed to confirm the existence of such a lesion. In keeping with the classic principle of clinical science, Finlay correlated symptoms with organic disorders. The endothelial lesion resulted in excessive filtration of the most fluid parts of the blood, leading to concentration and constriction of the vessels. Everything else followed: ischemia of the parenchyma, rupture of the capillaries, passive hemorrhages, and functional disturbances in various organs.

To stick to the essential, the interest of this pathogenic hypothesis lay in its dual function. On the one hand, the specific lesion indicated the probable seat of the morbid principle. On the other hand, it justified the drawing of an analogy between the transmission of yellow fever and that of another eruptive fever: "Assimilating the disease to small-pox and to vaccination, it occured to me that in order to inoculate yellow fever it would be necessary to pick out the inoculable material from within the blood vessel of the person who was to be inoculated."[29] In the behavior of the insects Finlay saw simply the old technique of inoculation.

If I were to summarize Finlay's perception, I would say that he saw the mosquito with Jenner's eyes. If I were to describe his behavior in the form of a paradox, I would say that it was because he saw the mosquito as a lancet that he opened a new chapter in medical entomology. To Finlay the proboscis of the mosquito was the equivalent of a stylet. But the mosquito was a *natural inoculative agent*. In order to play this role it was essential that it bite more than once: it had first to remove a substance, then inject it.

It should now be clear why Finlay distanced himself from Manson. Manson's problem was to confirm a hypothesis concerning the life cycle of filaria. Finlay's involved the transmission of an unknown principle. Manson focused attention on the life history of the nematode, Finlay on the transmission of the yellow fever germ. Finlay was thus forced to question what Manson took to be an established fact. Later the latter would confess: "A regrettable mistake, the result of want of books, was my belief that the mosquito died soon after laying her eggs."[30] In fact, it was not so much want of books as the nature of the problem he set himself that placed an obstacle in the way of Manson's seeking out or producing such information. By contrast, the problem that Finlay set himself led him to question nature, whereupon it turned out that nature was more than willing to provide him with evidence to confirm his hypothesis.

As we have seen, the mosquito entered the field of tropical medicine in the guise of intermediate host or, to use Manson's terminology, nurse. The mosquito aided the filariae in transit; it placed them in a position from which it was possible to gain access to their primary destination, access that coincided with their adult stage. The filariae were released into the water just as they became able to lead an independent life. Everything contributed to the appearance of a harmonious complex: observation of the adult forms, the apparently vivifying effect of water on the adult worms, and the return pathway to the human host via the drinking water. Manson had no reason to question the idea that the mosquito feeds on blood once and dies shortly thereafter. It seemed so natural to assume that the filariae were ingested with the drinking water that Manson anticipated the outcome of the critical experiment without

THE FORMATION OF A HYPOTHESIS

performing it. What if filaria-infested water were drunk by a human subject? "I am quite satisfied as to what would be the result."[31]

Finlay, by contrast, had two sound reasons for revising Manson's procedure. First, he abandoned the parasitic disease model. Indeed, it was for this reason that it became possible around 1880 to formulate a theory of transmission. In fact the parasitic disease model had to be abandoned because the yellow fever germ had yet to be identified: "I shall not concern myself with the nature or form of the morbific cause of yellow fever, beyond postulating the existence of a material transportable substance . . . something tangible which requires to be conveyed from the sick to the healthy before the disease can be propagated."[32] Second, the insect could function as a vehicle if and only if it made repeated bites. Unless the mosquito satisfied this condition, Finlay's hypothesis made no sense.

The differences between Finlay and Manson are now clear. Manson had demonstrated a new exit pathway for the filaria, but he relied on the "usual" entrance pathway, the drinking water. He saw the mosquito's bite primarily in terms of an advantage for the filaria, a means by which it could exit its human host and continue its life cycle. Thus he surbordinated the life cycle of the mosquito to that of the nematode. Finlay, on the other hand, enjoyed a wider latitude. Forced to set aside the problem of the yellow fever germ's natural history, he asked instead about its mode of propagation. Paradoxically, this limitation in terms of etiology was compensated by an expansion of possibilities in the realm of epidemiology. Finlay saw the transmission of the unknown germ as part of the mosquito's life cycle. He saw the mosquito's bite, moreover, in terms of the advantages it conferred on the insect. Thus the pathological agent was dissolved in a meal of blood. Further questions arose. Why did the female mosquito feed on blood? Why might she feed more than once? The answers to these questions contained the information Finlay needed to justify his hypothesis. That is why he took up his stand on the ground of natural history. Manson-Bahr would say of Finlay: "He had no real knowledge of entomology."[33] The imprudent fellow had not read the texts.

But Finlay knew plenty about entomology—more, in any case, than his contemporaries, who believed that after fertilization the female mosquito returned to the water to lay her eggs a short while before her death. What exactly did Finlay know? That the female captured in the act of fertilization and separated from the male consumed several meals of blood. From this came the decisive question: Was the blood meal connected with the insect's sustenance or with the process by which the eggs matured? So large a quantity of blood could not be for the purpose of feeding so small a body: "I have come to the conclusion that the sucking of blood is intended for another object connected with the propagation of the species. . . . If, for instance, the maturation of the ovules demands a temperature of $37°$ C., the latter could scarcely be obtained by any other means so readily as by the insect filling itself with a fair amount of blood of the temperature." The rest follows. If one blood meal sufficed for the maturation of all the female's eggs, then she would lay only once. If several blood meals were necessary, she would lay more than once. From this came a selection criterion: a single feeding and single laying implied a mosquito of large size; several feedings and several layings implied a mosquito of small size. Hence the agent of transmission of yellow fever must be a small mosquito.

This interpretive scheme or filter remained to be put to use. Finlay had a thorough knowledge of the island's insect life. He focused on two species of mosquito. One, the *zancudo* or *Culex cubensis*, measured five to seven millimeters from the root of the proboscis to the anal extremity. The other, which he designated *Culex mosquito*, came in two varieties: one large, slender, and the same size as *zancudo*;

the other smaller, just four millimeters. The next task was to compare the fertilization-feeding-laying cycle in *zancudo* and *Culex mosquito*. The zancudo and other large mosquitoes were able to take in sufficient blood in one feeding to permit the maturation of all 200 to 350 eggs, which were then laid all at once. By contrast, the smaller species needed to drink their fill of blood more than once before laying their eggs in several deposits: "Evidently, from the point of view that I am considering, the *Culex mosquito* is admirably adapted to convey from one person to another a disease that happens to be transmissible through the blood."[34] Nuttall claimed that Finlay had been content simply to choose the most common species. And Gorgas insisted that he chose *Culex mosquito* from among six or seven hundred other species. There are no grounds for either such dismissive snideness or such excessive praise.

Finlay was soon able to relate the geographical distribution of yellow fever to the geographical distribution of the mosquito. Where the disease was endemic, it broke out during the summer and vanished with the coming of cold weather. The seasons were either favorable or unfavorable to the insect's activity. In some cases epidemics died out quite close to the original source of infection. The insect's radius of action was limited; the *Culex mosquito* has small wings and is a poor flier. Since the nymphs are aquatic, moreover, the females do not stray far from water. In other cases, however, the disease was known to be communicated over great distances from the original source of infection. It was possible for the mosquitoes to hibernate in various crannies only to reawaken with the coming of spring. The existence of several vehicles accounted for outbreaks of the disease in European and American ports: hidden in clothing packed in a trunk, the mosquito could carry the germs of the disease in its lancets.

In the hope of clinching the mosquito's role in the

propagation of yellow fever, Finlay paid close attention to various peculiarities of their biology. His epidemiological investigations taught him about the insects' ecology, ethology, and chorology. Ecology, the science of relations between living things and the environment, covered such matters as optimal temperatures, resistance to temperature change, and seasonal emergence of the mosquito. Finlay also explored the nature of larval deposit sites. Ethology included studies of the mosquito's reproductive cycle and its relation to man. Finally, chorology, or the study of the geographical distribution of living things, concerned both the active displacement and the passive transport of the insect. Finlay thus identified various factors important to any understanding of the mosquito's role in the pathogenesis of the disease.

Nevertheless, Finlay knew full well that to convince his colleagues he would have to produce an experimental case of the disease. Such a project was impractical. Not only was the disease agent unknown, but experimentation on animals would have meant venturing onto shaky ground. A positive result would prove nothing in the absence of any criterion for deciding that the disease was indeed yellow fever. Neither would a negative result be convincing, since it was unknown whether animals were truly receptive. One course was still available: the method of Jenner. A single bite, Finlay believed, would cause a mild case of the disease and induce immunity. Out of this belief came several rules of practice. First, inoculations should not be made so close to an epidemic focus as to cast doubt on the results, nor should they be made so far away as to risk spreading the disease. Second, the mosquitoes used for the inoculations should be captured in places where the disease was unknown. In other words, the instrument used should be naturally "sterilized." Finally, the morbid material should be "collected" between the third and sixth day of the illness. The victim should be exposed to the contaminated mosquito as soon as the mosquito was apt to bite, that is, within two or three days. The method was appealing, but Finlay undoubtedly rushed ahead more quickly than he should have. Throughout the 1880s his dream was to preserve receptive individuals from the disease by administering a benign inoculation.

It was necessary to adopt the standpoint of historical epistemology in order to lay to rest a prejudice that has bedeviled the history of Cuban medicine, namely, that Finlay was unfamiliar with the work of Manson. No longer can Finlay's silence justify denial that he modified and then used Manson's theory. In so doing he came to see the mosquito as the vehicle of yellow fever. Finlay reconceptualized the meaning of this "object" by incorporating it into a new theoretical famework. In so doing he incontestably laid the foundations of medical entomology.

For twenty long years, however, his hypothesis would remain in obscurity. This fate was inevitable. What was needed to rescue Finlay's hypothesis from oblivion was Ross's work on malaria. In the meantime, however, bacteriology was all the rage. Finlay, like his contemporaries, therefore turned his attention to the search for the germ of yellow fever.

In 1880 Pasteur expected that his method of attenuating microbes would soon bear additional fruit, aiding humanity in its battle against disease. Research on yellow fever quickly focused on the search for a parasite. Physicians scoured "pathological sites" for the yellow fever germ. Early work on the disease was marked by the influence of various medical systems. Doctors were nevertheless quick to adopt the program of the new science, which they claimed shed new light on the problem. The first thing they would discover was the parasite. Next they would propose a pathogenic theory based on its properties. Then they would find a way to attenuate its virulence. This work demands our attention because it shows how contemporaries were mobilized by the transformation of scientific theory into a practical promise of salvation.

It was a strange encounter that took place between, on the one hand, enthusiastic but ill-prepared researchers and, on the other, scientists in possession of the latest techniques. New methods of prevention were alleged to have been found. One way to refute such claims was to show that the isolated microbes were not the cause of yellow fever. In the next-to-last decade of the nineteenth century French and American bacteriologists tested many alleged prophylactic vaccines. Numerous scientific missions visited Brazil, Mexico, and Cuba. Cultures were analyzed in well-equipped laboratories in Europe and the United States. Scientific research disproved the alleged discoveries.

In 1897 Havelburg and Sanarelli revived the debate by announcing that they had discovered the yellow fever germ. Although Havelburg's bacillus was soon forgotten, there was much discussion of the so-called bacillus icteriodes. The Italian physician's discovery caused a considerable stir. American doctors saw an opportunity to distinguish themselves. Some, like Sternberg, saw a chance to prove that they had anticipated Sanarelli's discovery, while others, like Wasdin and Geddings, hoped to confirm it. Sternberg soon withdrew his claims, however, after comparative studies discredited Sanarelli's bacillus. But Wasdin and Geddings insisted that the bacillus icteroides satisfied all of Koch's conditions. A raucous controversy ensued.

THE MICROBES OF YELLOW FEVER

Parasite theory had already scored its first successes: Pasteur had recently discovered vaccines against chicken cholera and anthrax. Medicine became conscious that a revolution was in progress. That revolution spread quickly to the New World. A vaccine maker who would later enjoy his moment in the limelight looked back on the hopes of that heady time: "The remarkable work of the parasite school, which opened up a new world in pathology, carried the faith into even the most recalcitrant minds, and those inspired by that work . . . would soon reap the rich fruits of their titanic efforts and struggles."[1]

The search for the yellow fever germ was the order of the day in the three countries where the disease was endemic: Brazil, Mexico, and Cuba. Doctors there felt compelled to unravel the mystery of the disease, to find its cause, to determine its nature and describe its evolution. The promptness of the response casts doubt on the scientific maturity of those involved, yet it reflects, in Finlay's words, the "humanitarian efforts undertaken in the three principle foci of this endemic disease to discover its etiology and prophylaxis."[2] The hope of producing an attenuated virus accounts for the speed with which these researches gave substance to the yellow fever germ. The new technique spurred dreams that soon took on reality in the New World. Not only would doctors find the etiological agent of the disease; they would also discover methods to prevent it and present optimistic statistics to prove their effectiveness.

Some researchers would use bacteriological methods to discover the germ of yellow fever. They sought to obtain pure cultures and inoculate animals to reproduce the disease. Others would follow a different course because what they were seeking was a fungus whose polymorphous nature ruled out the use of modern techniques. Surprisingly discrepant results were soon obtained. The discrepancies stemmed from lack of familiarity with the techniques of microbiology. One could choose to emphasize this lack of skill and the resulting errors, but to do so would obscure the reasons why different researchers were able to isolate germs that bore little resemblance to one another. Instead I shall stress what these different research programs had in common: all were based on the same principle of analogy, which underlay the variety of observations. Analogy revealed the nature of the pathogenic agent: the germ must be a microphyte because it behaved like a plant, as demonstrated by the fact that it flourished during the summer. By contrast, when the principle of analogy was turned into a law governing the production of species, it became possible to designate the parasites: the structure of the visible was based on the plant forms initially chosen as a model. After which the rest was easy: a pathogenic theory based on the properties of the germ, reproduction of the disease in animals, and conversion of the agent into a vaccine.

Four contradictory identifications were proposed during the 1880s. In Brazil Freire reported that the pathogenic agent was *Cryptococcus xanthogenicus*, while Lacerda singled out the fungus *cogumello*. In Mexico Carmona y Valle believed the specific agent was a mold, *Peronospora lutea*. In Cuba Finlay isolated a microorganism that he called *Micrococcus tetragenus febris flavae*.

Freire saw an analogy between the behavior of the yellow fever germ and that of plants: "The vegetative cycle, necessary to the full development of all plants, explains why the xanthogenic microbe's harmful effects declare themselves only at a certain time of the year."[3] Freire, however, "discovered" his microbe in a book: "Consider now what M. Robin says in his *Treatise on the Microscope*: 'The cryptococci constitute a cluster of ovoid cells. . . . The cryptococcus belongs to the malacophycaea subfamily of algae.' . . . *Cryptococcus xanthogenicus* therefore belongs either to the isocarpaean or malacophycaean family of algae."[4] What he saw confirmed what he had read. Freire described the growth and reproduction of the cryptococci. He found the germs in the urine, blood, and tissues. His pathogenic theory was based on the properties of the microbe, which acts like the anthrax germ. Inoculation of urine droplets sufficed to transmit both germs and disease. The next step was to attenuate the microbe. Freire believed that their energy increased during the hot season and that they fell into an inert state during the cold. Immunity could therefore be achieved without danger by inoculation with the culture during the winter. The emperor of Brazil backed this prophylactic technique. In 1890 Freire presented favorable statistics based on more than ten thousand cases.

Lacerda also noted an analogy between the yellow

fever germ, which flourishes only in the tropics, and exotic plants: "Do we not see in the vegetable . . . particular species become domiciled in certain zones . . . but once transported and placed in different surroundings, their vigor decays and they soon perish?"[5] Lacerda's a priori choice was a polymorphic fungus, which he then duly discovered in the bodies of patients: "The fungus *cogumello*, which is found in great abundance in the viscera and humors of subjects suffering from yellow fever, exhibits two quite distinct forms over the full course of its evolution."[6] The first—transitory—form is that of a mycelial tube that produces spores. During the second phase filaments appear, the ends of which sprout black masses known as sporules. When the germs enter the liver, they disrupt its functions. The kidneys are also invaded, leading to nephritis and anuria. Lacerda proposed a series of experiments involving the injection of substances derived from cultures of the fungus into the blood of animals. If these proved successful, a vaccine could then be produced.

Carmona y Valle also posited an analogy between the yellow fever germ and plants of alternating generations. He went so far as to compare the microbe's polymorphism to the metamorphoses of the tapeworm: "The zoospores are the tiny eggs of the *proglottis*. . . . [They] must undergo the mucedinous transformation outside the organism before reproducing yellow fever."[7] Carmona y Valle found the fungus he was looking for in the article on fungi in the *Dictionnaire Encyclopédique des Sciences Médicales*: "Bertillon . . . calls Peronosporaea those cryptogams . . . that exhibit oogonic dilations from which come zoosporanges laden with spores. . . . The plant that we are studying belongs . . . to the family Peronosporaea."[8] He then describes a type of fungus whose natural history was studied by Bertillon. Carmona y Valle recounts that history after his own fashion. The zoospores, which are found in large numbers in the urine, fuse and give rise to spores.

Expelled into the environment, these produce the mold, provided conditions are ripe for germination. This intermediate form in turn produces the fatal *Peronospora lutea*. A prophylactic method based on the fungus's polymorphism was then proposed. The harmless zoospores could be injected in the form of a vaccine made from the residue of dried urine diluted in distilled water. Carmona y Valle claimed to have successfully immunized several hundred individuals in this way.

Finlay had initially decided to leave the etiological question in suspense. He was forced to confront it, however, in 1883, when Corre raised an objection to his theory: the quantity of infectious material taken up by the mosquito seemed to him too small for a successful inoculation. Finlay sidestepped the difficulty: an individual might be bitten by one or more mosquitoes in which the active infectious substance had reproduced. Cooke and Berkeley argued that parasites developed while harbored by insect hosts: "We may therefore deduce that if a specific yellow fever microbe, bacterium, or fungus exists, it would find in the mosquito's lancet . . . sufficient space to grow and reproduce for as long as the diptera continues to live."[9] This deduction, based on an analogy, was one of two responses Finlay gave to Corre's objection.

It was also the point of departure for a line of research that would lead to discovery of the specific germ. A few months later Finlay came to believe that the culprit must be a fungus. He had just learned that Grawitz and Leber had proposed distinguishing bacteria from fungi by the criterion that fungi exhibit noxious effects when introduced into an organism by means of inoculation. In 1884 Finlay claimed to have found fungi in the probosci and saliva of mosquitoes. In 1886 Finlay and Delgado found colonies of fungi in their cultures. In the following year they detailed the character of the microbe and associated it with its prototype, Gaffky's *Micrococcus tetragenus*: "We regard the success of these solid-medium cultures as the first firm and decisive step in microbiological research on yellow fever."[10] Finlay subsequently accumulated proofs of his hypothesis. He found his microbes in both inoculated animals and mosquitoes. And he of course produced statistics demonstrating the efficacy of his method of prevention.

Four etiological agents were identified. These divergent results soon gave rise to violent polemics. It was alleged by some that the morphological changes that Freire claimed to have seen in his microbes were completely fanciful. Lacerda, it was charged by others, had adduced impossible forms to support his claim that his fungus was polymorphous. Everyone agreed that Carmona y Valle had mistaken a variety of germs accidentally introduced into his test tubes for a single plant. The details of these polemics may be safely ignored. The essential point is this: that etiological agents were proposed on the basis of vague analogies and imaginary pathogenic processes. The pathological agent was in each case the centerpiece of a medical system. And each researcher dreamed of distinguishing himself as the Pasteur of the tropics.

Where the physicians involved saw definitive answers, a detached observer would have seen open questions. Which of these several germs was the specific agent? Had not these New World researchers gone wrong somewhere? Very likely they had. Yet their vaccines were being heralded at a time when yellow fever posed an obstacle to construction of the Panama Canal and when it was liable to break out at any time in American ports. The claims had to be examined. Thus the etiological issue was subject to the scrutiny of bacteriology.

EFFORTS OF VERIFICATION

The results obtained in Latin America seemed too good to be true. When Freire's work came to the attention of the

French Academy of Medicine in 1884, Rochard had this comment: "The scientists of Rio de Janeiro are truly privileged. Nature reveals to them all her secrets with an obligingness she has never shown to Old World researchers. Their discoveries are always complete; after them, nothing remains to be found out."[11] Boulay's response was more indulgent: "I do not wish to vouch for the experiments of Mr. Domingos Freire, but I do feel that we should be careful, lest by criticizing them too severely we may discourage attempts similar to that recently made by the physician from Rio."[12]

In the United States the confrontation took a more violent turn. Progressives, who favored vaccination, seemed to clash with conservatives over public health issues. But the real bone of contention was the budget: some favored free spending, while others wished to economize. Holt, speaking before the American Public Health Association, argued in favor of sending a costly investigating commission to South America; he asked for an allocation of $30,000. Meanwhile, Caldwell derided the work of the Latin American physicians. His purpose, he said, was "to defeat the creation of a useless commission and prevent the incurring of a needless expenditure of public money."[13] The medical press was divided. Ultimately a compromise was reached, one that was both economical and discreet. It was decided that a single expert would be sent to examine the discoveries of the Latin doctors.

Thus scrutiny came from two quarters: France and the American Dr. Sternberg. Gibier was dispatched to Cuba in 1887 and to Florida the following year. Aristide Le Dantec, serving in French Guiana, had the opportunity to observe several epidemics, out of which came his thesis in medicine, *Recherches sur la fièvre jaune* (1886), which treated the etiology of the disease. Vincent, while on a research mission to study the American coast, met Finlay and Carmona y Valle. In Paris, Cornil and Babès examined samples of cultures sent by Lacerda. This work may be seen as an offshoot of a larger research effort conducted during the 1880s. Recall that Monard and Talmy had brought a variety of preparations from Senegal back to France. Capitan and Charrin examined Monard's samples and found microbes. Talmy's samples, however, yielded negative results.

In the United States the verification of claims concerning yellow fever was entrusted to the eminent Dr. Sternberg, upon whom Koch would later bestow the title "father of American bacteriology." He received his first orders, signed by Grover Cleveland, on April 29, 1887. His mission was to examine the value of the various methods of inoculation and the whole etiological issue. During his time in Brazil, however, the yellow fever lay dormant. He therefore undertook a further mission to Cuba (special order no. 93 dated April 23, 1888). This time his visit was too short, as he was forced to file his report before June 25. At his own request Sternberg was then sent to Decatur, Alabama, which had been stricken by an epidemic (special order no. 224 dated September 26, 1888). Finally, Sternberg was again sent to Cuba, on this occasion without a time limit (special order no. 30 dated February 5, 1889). In Havana he obtained abundant samples, which he examined in his laboratory at Johns Hopkins. In 1890 he published his voluminous *Report on the Etiology and Prevention of Yellow Fever*, a study that closely followed the pattern laid down by the Chaillé Commission ten years earlier.

The purpose of all these investigations was to evaluate the announced discoveries. The bacteriologists were confronted with an alternative: did the samples test positive or negative? At first the results were all negative. Organ and blood samples revealed no trace of the microorganisms described by the Latin physicians. What, then, was the source of the confusion? Numerous pitfalls awaited anyone bold enough to seek the specific agent of yellow fever. Most commonly, observers mistook pigmentary granules, deformed blood cells, or tissue fragments for organic forms. Freire, Lacerda, and Carmona y Valle all fell into such traps. All the experts agreed: not a single microbe had turned up. All decried the same errors. *Micrococcus xanthogenicus*? "Dr. Freire has mistaken deformed blood corpuscles, fat globules from the liver, and the debris of tissue elements . . . for microorganisms."[14] The fungus *cogumello*? "Monsieur Cornil and I are convinced that Monsieur Lacerda was mistaken and that he described as parasites filaments of plant tissue and pigment."[15] *Peronospora lutea*? "Monsieur Carmona's alleged microorganisms are nothing other than granules of one sort or another subject to Brownian motion."[16]

Then positive results began to turn up. Microscopic examination revealed the alleged microorganism. Was the microbe the specific agent, however? If it was found not to be present in some victims of the disease (or present in nonvictims), it could be safely dismissed. What, then, accounted for the error? In most cases the experimenter believed that he was culturing a specific agent and following its evolution. Unbeknownst to the observer, however, foreign parasites had taken hold. Finlay was caught in this kind of error. Sternberg found the micrococcus he had isolated, but its presence was accidental. The microbe had been introduced when blood was drawn from patients: "My researches show that it is a very common organism upon the surface of the body of patients . . . and quite as common in cases not having yellow fever as in those sick with this disease."[17]

In their investigations bacteriologists soon discovered other germs. They thus substituted new microbes for those they had just discredited. Babès made a discovery in the material sent by Lacerda. The blood contained an enormous quantity of parasites: "They are tiny diplococci, in some cases linked together in short chains composed of four individual organisms."[18] Gibier, meanwhile, isolated a gelatine-liquefying microorganism from the contents of the intestine. It was a slender, not very refractive bacillus similar to the vibrio of cholera. When the intestines of guinea pigs were inoculated with a culture of this microbe, the animals died. In the following year Gibier found his bacillus in three out of seven disease victims. He was convinced: "I think I am right in saying that the presumption that this bacillus is the cause of yellow fever tends to become a certainty."[19] Finally, Sternberg found the bacillus *x*, to which he was tempted to assign a role. The microbe occurred in the alimentary canal; cold did not destroy its virulence; and it was pathogenic in rabbits: "It is possible that this bacillus is concerned in the etiology of yellow fever."[20]

These new candidates were quickly ruled out, however. Babès was soon to abandon the notion that the bacillus he discovered had anything to do with yellow fever. After examining specimens from three cases he concluded: "Despite the most scrupulous research and the advice of Koch, it proved impossible to discover any chains in the brain, kidneys, liver, or spleen. . . . The question of whether these microorganisms are truly the cause of the disease . . . therefore remains unresolved."[21] Sternberg found Gibier's bacillus in the contents of certain intestines, but in most it was not present. He therefore concluded that it was an accidental microbe: "My extended researches give no support to the suppostion that it is concerned in the etiology of yellow fever."[22] As for bacillus *x*, Sternberg found it in only half the cases he examined, and he therefore dismissed it as a possible cause: "This bacillus . . . has not been found in such numbers as to warrant the inference that it is the veritable infectious agent."[23]

Research thus demonstrated that none of the microorganisms found could be viewed as the specific agent of the disease. The vaccines had no prophylactic value. The South American doctors' contribution was to have engaged the interest of experts in the disease. The *Annales de l'Institut Pasteur* had several times been asked to offer an opinion of their work. After the results of Sternberg's investigation were made public, the journal's editors broke their prudent silence and published a report on the American bacteriologist's research: "Although no positive results were obtained in regard to the etiology of the disease, this work deserves credit for having . . . demonstrated the difficulty of the question."[24] The usual bacteriological methods had proved inadequate to discover the yellow fever germ, but no thought was given to scrapping the methods. Research would have to continue.

SANARELLI'S BACILLUS

In 1890 Sternberg showed that none of the microorganisms considered over the past ten years was the cause of the disease. His negative conclusions bolstered the notion that the agent was to be found in the digestive apparatus. Sternberg often stated that the yellow fever germ, like the cholera germ, probably lived in the intestine. In 1897 Havelburg confirmed this analogy: "Yellow fever is a disease whose specific toxic agent enters the stomach and develops there as well as in the intestines. . . . This is analogous to what takes place in Asiatic cholera."[25] Injecting the blood of a yellow fever victim into the veins of an animal sufficed to demonstrate the existence of a toxic substance. The etiological agent that produced it, however, lived in the digestive tract. Havelburg found it there in large numbers—and without great difficulty, since it was the only organism present. He described a small, thin bacillus resembling a diplococcus and belonging to the group of colonic bacilli. Optionally anaerobic, it was easily stained. What is more, it was pathogenic in guinea pigs.

In the same year Sanarelli announced the discovery of the bacillus icteroides. In order to understand his successful work at Montevideo's Institute of Health, however, we must first consider work done at the Institut Pasteur, initially in Roux's laboratory. Specifically, I am referring to the work of Roux and Yersin on diphtheria. They found the bacillus in the false membranes and showed that a culture at one point in the body gave rise to a general infection. From this they developed the idea that the microbe secretes a highly active toxin, whose effects they demonstrated. Then, in Metchnikoff's laboratory, Sanarelli did work on the pathogenesis of typhoid fever. His work was inspired by that of Roux and Yersin. But clinical observations, pathological anatomy, and bacteriological considerations led him to reject the idea of an infectious process of local origin. Abdominal symptoms were not the first sign of the disease, nor were they invariably present. Lesions were sometimes benign or nonexistent. Eberth's bacillus was rarely found in the digestive tract or the feces of typhoid victims. For these reasons Sanarelli came to the conclusion that typhoid fever was a general infection produced by a poison. The toxin's effects on the intestine opened the way to secondary invasions.

When he came to study yellow fever, Sanarelli based his view of the pathogenic process on that of typhoid: "Yellow fever must belong to the typhic group of diseases. It is a febrile disease, essentially toxic, whose most serious and important complications are by no means limited to the digestive tract."[26] Yellow fever was not a disease whose seat lay in the alimentary canal. The infectious agent, moreover, did not secrete toxic substances into the digestive apparatus. All the evidence seemed to confirm these conclusions: gastrointestinal symptoms in some cases were mild or even nonexistent; the severity of intestinal lesions varied; and the specific bacillus was never found in the intestinal contents or feces of yellow fever victims. The etiological agent was probably located in the circulatory system. Sanarelli believed that heptic steatosis, nephritis, vomiting, and hemorrhage were all effects of the toxin. The toxin's effects on the digestive tract facilitated secondary infections by colonic bacilli, streptococcus, and staphylococcus leading to septicemia, the results of which were sometimes fatal.

Sanarelli claimed that this pathogenic theory grew out of his bacteriological research. In fact it was the other way around. It was because he believed that the pathogenesis of yellow fever resembled that of typhoid fever that he soon found the specific agent in several specimens of blood and kidneys: "These fragments were found, post mortem, to contain an abundant proliferation of specific microbes, exactly like that known to occur in the splenic pulp of typhics."[27] Sanarelli found his bacillus in half the cases he examined and cultured it in the usual media.

At the time Thoinot wrote: "The great question of the etiology of yellow fever, which for a time seemed to have been abandoned, has been raised anew by the work of Sanarelli and Havelburg."[28] In 1897 the two researchers proposed two specific bacilli. One was a parasite always found in the digestive tract; the other occurred in the blood, liver, and kidneys. Which was the true yellow fever microbe? Or was neither responsible for the disease? Thus two more microorganisms were added to the already long list of microbial flora commonly associated with yellow fever. The Institut Pasteur soon received specimen cultures. Under Roux's direction Novy embarked on a comparative study of the two bacilli. Havelburg's microbe belonged to the group of colon bacilli, Sanarelli's to the group of typhoid bacilli. "Neither the Havelburg nor the Sanarelli bacillus is the cause of yellow fever." Novy, who was familiar with the work of Roux and Nocard on the agent of pleuro-pneumonia, suggested that the yellow fever microbe might belong to the group "of organisms smaller than the "infinitely small bacteria."[29]

Novy's criticisms struck down Havelburg's theory but did no damage to Sanarelli's reputation. Why was the fate of the Italian physician's work different from that of Havelburg's? Why did the bacillus icteroides become the focus of a bitter polemic? That polemic seemed to point up uncertainties inherent in bacteriological verifications. It seemed to show, as Warner put it, "that the ideal guidelines established by Koch were open to error and disputation at practically every point."[30] The scientific debate was complicated by a clash between two medical institutions: the Army Medical Corps and the Marine Hospital Service. The army doctors rejected Sanarelli's discovery in light of comparative research ordered by Sternberg; but under Wyman's command the Marine Hospital Service took up the Italian physician's cause.

The divergence of interpretations and confrontation of institutions were, however, mere surface effects. In reality, the controversy took place because Sanarelli's discovery provided American medicine with an opportunity to distinguish itself. By linking the future of the bacillus icteroides to that of his bacillus x, Sternberg delayed verification and confused an issue that should have been fairly quickly resolved at the empirical level. By attempting to prove that the bacillus icteroides was the specific agent of yellow fever, the Marine Hospital Service doctors fueled the controversy and complicated efforts of verification. If it were necessary to apportion blame, I would absolve Sanarelli: his only mistake was to have believed in his discovery. By contrast, for three years American medicine seized on the bacillus icteroides as a pretext. With the controversy it paid a high price for its determination to establish American preeminence in the field of bacteriology.

If Sanarelli's research carried conviction, it also left a number of points in need of clarification. The discovery provided American medicine with an opportunity to shine. An American's name appeared alongside that of the celebrated Italian—alongside and ahead of, if one believed, as Sternberg did, that the bacillus icteroides and bacillus x were one and the same: "If my bacillus x and the bacillus icteroides of Sanarelli are proved to be identical . . . Sanarelli will deserve full credit for the demonstration."[31] Credit for the *discovery*, therefore, would belong to Sternberg. American names would come after the Italian's, however, if one believed, along with Wasdin and Geddings, that their work had confirmed Sanarelli's: "Our commission had proved what Sanarelli did not, that the bacillus icteroides is constantly present in yellow fever, and is the cause of the disease."[32] Hence they deserved credit for providing proof.

Sternberg's initial concern was to assure his priority. Upon learning of Samarelli's research he immediately took a position. In September 1897 he exhumed his bacillus x and noted resemblances that argued in favor of the identity of the two microbes. Sternberg implied that Sanarelli's work was, if not plagiarism, at least an extension of his own. Proof was needed, however. Comparative study was the only way to resolve the question. Even if Sternberg linked the bacillus icteroides to his bacillus x, the problem remained open to discussion. Either the two were identical, in which case the yellow fever germ would be called the Sternberg-Sanarelli bacillus, or they were different, in which case it was back to square one. The first verdict was handed down in 1898. Reed and Carroll, who had conducted a comparative examination of the two bacilli, reported their conclusions: "Bacillus icteroides (Sanarelli) is a variety of the hog-cholera bacillus and . . . should be considered only as a secondary invader in yellow fever." Bacillus x was not the same as bacillus icteroides: "It will suffice here to state our opinion that bacillus x should be placed with the colon group."[33]

Sanarelli, meanwhile, allowed his disciples the privilege of confirming his discovery. Bacillus icteroides did not yet satisfy all the conditions laid down by Koch. The first requirement was to find significant numbers of the microbe. On this point Sanarelli's work left something to be desired. In 1898 Wasdin and Geddings found bacillus icteroides in thirteen out of fourteen cases, with only one negative case. Sanarelli had tested the pathogenic powers of the bacillus on animals, but his research was incomplete. The Americans had therefore set out to do comparative studies. Their success was complete: "The bacillus icteroides fulfills the third postulate of Koch, and necropsies in some animals meet the further demand of the fourth, that the organism must be found generally distributed, as in the human body from this disease."[34]

In 1900 Sanarelli may have believed that he had emerged victorious from the controversy triggered by Reed and Carroll's results of April 1898. He could point to the numerous corroborative findings obtained by his disciples. The latter had found, isolated, cultivated, and inoculated bacillus icteroides. But even as Sanarelli was painting a flattering portrait, Agramonte reported the results of research he had begun in December 1898: "The specific pathogenic microorganism of yellow fever is as yet an unknown entity."[35] It was not until the summer of 1900, however, that the commission dispatched to Havana by Sternberg definitively ruled out the bacillus icteroides as the cause of yellow fever. The commission studied eighteen cases of the disease. Blood, drawn with the ususal precautions, was tranferred to sterilized test tubes. The bacillus icteroides was not present in the cultures examined. Post mortem investigation of the blood and organs of eleven subjects also proved negative. Hence "Bacillus icteroides (Sanarelli) stands in no causative relation to yellow fever, but when present should be considered as a secondary invader in this disease."[36]

The Pasteurian theme dominated the last two decades of the nineteenth century. Ultimately, however, the meandering path of bacteriology led nowhere. At the same time the American commission reported its negative findings, it also announced that the mosquito selected by Finlay was the intermediate host of the yellow fever germ. Having come to Cuba to confirm or disconfirm Sanarelli's discovery, the commission returned home with the host of an elusive parasite. But that is another story.

The Theory Verified

It took twenty years before anyone evinced the slightest interest in Finlay's mosquito. In order to explain the impenetrability of the problem, the first thing we must do is show how it took a transformation of the field of parasitology to open up the epidemiological line of research. Just as Manson's work enabled us to understand Finlay's procedure, so will Ross's work enable us to grasp the procedures of the American investigators. The names Laveran and Ross will serve as beacons marking the period from 1880 to 1898, the time that it took researchers to discover the fruitfulness of the work done on the transmission of malaria.

Oddly enough, the American Yellow Fever Commission landed in Havana in late June without the least thought of the work of Ross or Finlay. In fact it was initially occupied by the search for Sanarelli's bacillus. It was not until late June that it was set on the right path when the Liverpool commission brought up the mosquito theory. That is why Lazear conducted the first inoculations, in August, in such haste. We shall therefore be obliged to lay another legend to rest: that the American commission was solely responsible for the triumph.

Finally, we shall see that Reed's experiments, the culmination of the commission's work, were in fact quite routine. His results of the winter of 1900 simply confirmed those obtained several months earlier by Lazear.

We begin by tracing the process that finally brought the question of tropical diseases into clear focus and made it possible to understand how yellow fever is transmitted. It would be tedious to recount this history in detail. For nearly twenty years, from 1880, when Laveran's paper "On a New Parasite Found in the Blood of Several Patients Afflicted with Marsh Fever" appeared, to 1898, when Ross gave the solution to the problem, malaria was a major focus of medical research.

In 1880 Laveran discovered the hematozoon responsible for malaria. He described an amoeboid parasite that lived in the victim's blood cells. Over the next few years Golgi combined various of Laveran's observation to produce a picture of the so-called schizogonic cycle. This and much other research resulted in the formulation of a number of important laws governing this group of organisms. The parasite was a sporozoon of the coccidium family. Organisms of this family multiply in the blood by sporulation, with attacks of fever occurring when the spores are liberated. At least three varieties of the parasite were found in man. One produced quartan fever, another tertian fever, and the third irregular (pernicious or estivo-autumnal) fever.

Laveran's work and that of his earliest followers left two problems unresolved. First, what was the nature and role of certain bodies that were observed to emit long, mobile filaments? When such forms were observed in extravasated blood, should they be taken to indicate a disintegrating parasite or a new stage of development? Grassi and Bignami believed that the phenomenon was one of simple degeneration. Laveran, Golgi, Danilewsky, and Pfeiffer believed that these filiform bodies played a role in the parasite's transition from its intrabodily to its extrabodily stage. The second unresolved problem was that of the disease's propagation. Malaria had been produced by inoculating healthy individuals with blood taken from victims of the disease. Apart from this artificial means of transmission, the affliction did not appear to be directly communicable. Did the parasites live as saprophytes in stagnant water? Did they enter the organism when infected water was ingested? Or when dust from dried swamps was inhaled? Other pathways were also suspected. Mendini and Bignami took up King's old idea that the parasite infected mosquitoes, which than carried it from swamp to man. Conversely, Laveran and Koch argued that the mosquito carried the parasite from man to swamp.

In 1894 Manson took his filaria hypothesis and applied it to the mobile filaments, thus linking the two problems. The filaments appeared only in extravasated blood, hence they must be true flagellated spores, the first stage of the parasite's life outside the human body. What was observed in vitro must also take place inside the body of an insect: "The mosquito having been shown to be the agent by which the filaria is removed from the human blood vessels, this, or a similar suctorial insect must be the agent which removes from the human blood vessels those forms of the malaria organisms."[1] Following the death of the mosquito, the parasites presumably entered the water and from there infected man either by way of the drinking water or via the old mechanism of an atmospheric miasma.

This working hypothesis suggested an experimental protocol. Allow mosquitoes to bite victims of the disease whose blood contained cells apt to produce the flagellated bodies. Clearly, the stomach of the insect was the place to look for those bodies. The mobile filaments could be followed from there to their destination in the body of the host. Observations would have to be carried out on various mosquitoes until the right species was found. The hypothesis did not rule out the possibility of a cycle different from the one envisioned by Manson, nor did it rule out the possibility of an infection pathway other than drinking water or aerial miasma. The hypothesis certainly contributed to the solution of the problem.

That solution emerged from observations made by Ross—not without difficulty. The object of study was infinitesimal and troublesome: the extreme delicacy of the filaments complicated observation, and nothing was known about the shape or location of the object being sought. Nor was it certain that the right species was being observed. Ross came up with several strategies to circumvent these difficulties. First, he experimented on sterile mosquitoes, that is, mosquitoes hatched from larvae in captivity. This enabled him to avoid complications stemming from the intrusion of foreign bodies. Second, he was keenly aware of the variety of forms that the flagellated bodies might take thanks to his familiarity with insect parasites and with the normal and pathological histology of the mosquito.

During the summer of 1897 Ross made the key observation on two spotted-wing mosquitoes obtained from larvae in captivity. The insects had been allowed to feed on patients four and five days prior to observation. In the thickness of the stomach wall Ross discovered "pigmented cells." This observation revealed the location and appearance of the parasites within the insect's body. At around the same time, MacCallum discovered the biological significance of the phenomenon. He observed, in vitro, the crow hematozoon and identified hyaline and granular bodies. The hyaline bodies were observed to penetrate the granular ones: "We can thus consider the two forms of adult organism found in the blood as male and female."[2]

Manson reconciled the process described by MacCallum with the observations made by Ross. The "pigmented cells" were no doubt sporocysts lodged in the mosquito's body. Following Manson's advice, Ross turned his attention to aviary malaria. The development of the pigmented cells still remained to be traced; their final position in the insect's body promised to yield the key to the mode of infection. Ross first exposed birds with and without *Proteosoma* to the bite of gray mosquitoes. He found pigmented cells in the mosquitoes that had fed on birds with the parasite but not in the others. Therefore the pigmented cells in the mosquitoes derived from the parasites. He then conducted dissections over a period of several days. Until the eighth day the cells increased in size. They then exploded, liberating the filiform bodies. "Then came a step in the investigation of great consequence: it was none other than the discovery of those vermicules in the venenosalivary glands of the mosquito."[3]

We may ignore the work of Smith and Kilborne on the transmission of *Pirosoma bigeminum*, the cause of Texas cattle fever, by ticks, as well as Bruce's work on the transmission of *Trypanosoma nagana* by flies. In neither case did the work resolve the problem of the parasites' evolution in the vector's body. We may also ignore the polemic with the Italian school. The discovery of *Proteosoma*'s evolution in the mosquito applied to other parasites of the same group. The crucial point is that Ross effected the conceptual conversion for which Manson's method had paved the way but which it had not completed. This experimental structure would come to dominate exotic pathology. Research on filariasis would soon be adapted to fit this framework.

The analogy between the *Plasmodium* and the filaria had made it possible to explain how malaria is transmitted. In return, this new understanding would soon lead to the discovery of a new stage of development in *Filaria bancrofti* and thus settle the question of how filariasis is transmitted. Thomas Bancroft had pointed out the weak points in Manson's theory, raising some doubt as to its validity. For one thing, the mosquitoes lived longer than Manson had assumed. Believing that the final blood meal was the insect's only nourishment, he had neglected to feed his mosquitoes. What is more, filariae died rapidly after being placed in water: "Water therefore cannot be the medium, as was generally supposed, by which they ultimately reach the human subject."[4] Low, however, succeeded in establishing the filaria's life cycle. He examined sections made at different stages in the mosquito's life. Around the twentieth day the filariae penetrated into the sheath of the proboscis: "The mosquito bite is the natural and, probably, the constant and only opportunity for the filaria to enter upon its next and final stage of development in the definitive host."[5]

Thus mosquitoes had been shown to play a role in the transmission of malaria and filariasis. It was plausible to think that they played a similar role in the transmission of yellow fever. The logic of the history of scientific ideas would suggest an analogy, which was obvious to anyone working in the field of tropical medicine. Indeed, medical history had already bestowed its honors on one particular individual. As early as July 1900 (!) Guiart was able to write: "Mosquitoes are the agents in the transmission of two terrible tropical afflictions, filariasis and malaria. It seems likely that they are also responsible for the transmission of that dreadful scourge known as *vomito negro* or yellow fever. . . . Charles Finlay was the first to believe that mosquitoes might play a role in the transmission of the disease. . . . According to him, the intermediary was probably the *Culex mosquito*."[6]

The following month, the first results were obtained, albeit in tragic circumstances. Lazear took the mosquito provided by Finlay and coupled it with the theory provided by Ross to produce the first experimental case of yellow fever. What seemed obvious to Guiart, however, had yet to occur to the head of the American commission in Cuba, Walter Reed. What would soon be portrayed as the great triumph of American medicine did not come about without hesitations and errors over which American historians have drawn a discreet veil. What exactly took place in Havana during the summer of 1900?

THE PROOF

The conditions for a successful demonstration were all in place: Ross's theory, Finlay's idea, the right species identified. The program of research carried out by the American commission in the summer of 1900 was thus something that was "in the air." Yet three curious facts immediately stand out. First, the American commission did not initiate that research program directly after its arrival because bacteriological investigation was its first priority. Second, just when the decision to change direction was finally taken in early August, the commission's chairman was recalled to Washington. Finally, the first experiments were carried out so hastily that their results convinced no one. Apparently these facts have embarassed American historians, whose accounts are so highly romanticized that the crux of the story has been lost.

Let us therefore start from the beginning. We shall identify various strands and weave them into a new account of the events in question. No doubt the work of scientists is governed by the logic of the history of scientific ideas. The phenomenon we are dealing with here, moreover, transcends the individuals involved. Yet history is not a story dictated by Fate. This transindividual phenomenon must be described in such a way as to bring out the responsibility of individual human beings. In order to describe the "linkage of events," we must take a microscopic approach to the history of science. When we do, we arrive at the surprising conclusion that the credit usually bestowed on Reed must be reapportioned among Durham, Myers, and

Lazear—Durham and Myers being responsible for the decision to pursue the mosquito theory and Lazear for the proof of its validity.

Very soon after the event American historians insisted that the decision to examine the mosquito theory had been made by the army, either by the medical chief, Sternberg, or by the chairman of the Yellow Fever Commission, Reed. The historians in question based their accounts on the testimony of those who participated, either intimately or from afar, in the adventure. Truby's account was based on Sternberg's 1901 statement that he had suspected that the transmission of yellow fever, like that of malaria, required an intermediate host: "I therefore suggested to Dr. Reed . . . that he should give special attention to the possibility of transmission by some insect."[7] Torney and Owen, on the other hand, based their version on Agramonte's testimony, which related to an epidemiological investigation at Pinar del Rio. There, around July 20, he and Reed confronted the same phenomena observed by all who had studied the disease: its noncontagious character, the apparent harmlessness of *fomites* (articles contaminated by disease victims), and the possibility of remote infection. "It was there that for the first time the probability of mosquito agency in transmitting the disease was seriously discussed by members of the commission, and it was decided to carry out some research in this direction."[8]

These reconstructions are historically false because they are based on false testimony. It is understandable that Sternberg should have wished to claim credit for the decision himself: after twenty years of intensive but fruitless effort, he saw his subordinate covered with laurels. But Sternberg never took the mosquito theory seriously. The order that he issued, moreover, reflected his old passion, recently rekindled by Sanarelli's discovery: "It is evident that the most important question which occupies your attention is that which relates to the etiology of this disease."[9] And Agramonte's claims simply are not credible. When he mentions such extremely well-known observations, is he not reversing the actual order of things? It was the mosquito theory that constituted an interpretive scheme: from theory to observation, not the reverse. Furthermore, we know that what Reed and Agramonte were looking for at Pinar del Rio was Sanarelli's bacillus, as is indicated by this communication from Reed, dated July 24, 1900: "We have been able to study carefully seven cases since our arrival. . . . Neither during life nor after death have we been able to isolate Bacillus icteroides, but our sixth autopsy, which occured day before yesterday, cannot be definitely reported upon as yet."[10]

These falsifications are significant in the history of American medicine: they serve as retrospective justifications. Once the mosquito theory was proven, it became necessary to provide a transfigured history in which American medicine claimed responsibility for the decision. All these accounts are consequently haunted by a specific memory, which is invariably repressed: namely, that the Americans decided to investigate the mosquito theory at the behest of the Liverpool commission, which visited Havana in the second half of July. The English guidance could not be avowed for obvious reasons. For Sternberg, the commission's founder, it was difficult to admit that he had not divined what the English had been so quick to notice. For the American commission, which had covered itself with glory as the conqueror of yellow fever, it was difficult to admit that it had espoused the triumphant cause as the result of a suggestion made by others. And for American historiography, which has glorified this adventure, it was difficult to admit that the heroes of the tale were so distracted that they failed to notice in July what they would demonstrate the very next month.

The story was therefore cleaned up. It was a simple matter to conceal the English commission's key role. According to one version, the American commission went to Cuba with orders in hand ready to be carried out: "Consequently the members of the commission made Dr. Finlay a visit . . . late in June or early in July, and brought back mosquito and eggs." The danger was thus avoided. Not only did the English play no part in the decision, it was they who needed to be educated: "Reed took them through his laboratory and freely outlined the commission's work including the plans for testing the mosquito theory. . . . On a large table in the center of the room there were several glass laboratories containing mosquitoes."[11] But Sternberg never issued any such order; the visit to Finlay did not take place in late June or early July; and the mosquitoes that Reed showed the English were the ones Lazear had captured for his work on malaria.

According to a second version of the events, the commission allegedly decided to explore the mosquito thory in the wake of the Pinar del Rio investigation. Hence the visit to Finlay can be situated, correctly, in early August. Nevertheless, the role of the English commission is still concealed: "The illumination was sudden; it was almost meteoric. . . . On his return to Havana from Pinar del Rio, Reed counseled the commission to leave the bacterial trial and concentrate on a search for the agent of diffusion."[12]

In truth, the men who landed in Cuba were exhausted by three years of fruitless bacteriological research. Competent microbiologists, they had little knowledge of the epidemiology of yellow fever or malaria. Except for Lazear, the commission members were subject to Sternberg's stern control. They followed his order without hesitation—much to Lazear's annoyance: "Reed and Carrol . . . are interested in the controversy with Sanarelli and think of that all that time."[13] Howard would later recount his conversation with Reed: "He went on to tell me that he had been doing some work at Johns Hopkins University and had often talked with W. S. Thayer who had visited

Italy and studied the Anopheles-malaria work there and had therefore become anxious to do experimental work with yellow fever and mosquitoes."[14] Here we see the earnest plea of a man who knew full well that he was engaged in work for which he was not prepared. Reed stated that he had become familiar with the subject of diseases transmitted by arthropods. Why did he go to Finlay in search of a mosquito which, incidentally, he was unable to identify, since he immediately sent it to Howard? Reed said he was impatient to do experimental work. Why did he delay his decision until August?

In order to clarify this history, let us first set forth the facts. On June 25 the commission held its first meeting. Tasks were assigned in view of Sternberg's order to investigate bacillus icteroides. For the next month research continued in this direction. In mid-July, however, the Liverpool commission visited Havana. The historian Bean mentions a "breakfast in honor of the English." He reports the dishes that were served. The anecdote is pleasant, but Bean says nothing about what was discussed. Yet we know that information was exchanged. In an article published in the September 8, 1900, issue of the *British Medical Journal* (before the Americans announced their results) Durham and Myers gave their impressions of the trip. In Havana Carter had informed them that roughly two weeks—the "period of extrinsic incubation"—elapsed between the observation of an initial case of the disease and the emergence of a second case. Reed told them that two nurses who had cared for one patient contracted yellow fever two weeks after the patient died. Durham and Myers *connected* Reed's information with Carter's observation, mentioned Ross's theory, and rediscovered Finlay's hypothesis. Association of ideas made the conclusion inescapable: "Some means of transmission by the aid of an intermediate host— a town-loving host for this town-loving disease—is to some extent more plausible than might be anticipated."[15]

THE THEORY VERIFIED

Immediately after the departure of the English, the American commission met for the second time. "The final determination to investigate the mosquito theory was arrived at during an informal meeting of the commission . . . early in August 1900."[16] Duties were therefore reassigned: Carroll and Agramonte would carry on with bacteriological work, but Lazear would take up the mosquito theory. At this point the commission had its famous meeting with Finlay, from which the Americans came away with eggs of his *Culex* mosquito. A specimen was immediately sent to Howard for identification: it was the *Culex fasciatus fabricus*. But then, in a dramatic turn of events, the commission lost its head just when it received the mosquito eggs. In early August Reed was recalled to Washington to put the finishing touches on a report on typhoid fever at military bases, because he also chaired the commission named to look into that issue two years earlier. To sum up, the English left Havana at the end of July, and the Americans decided to examine the mosquito theory in early August. Finlay handed over his eggs, and Reed swiftly left Havana. In due course the eggs hatched. In summer it takes ten days for the mosquito to mature to adulthood.

Lazear therefore began his experiments on August 11. Initially he recorded nine negative cases. The mosquitoes were allowed to feed on yellow fever victims after the fifth day of sickness and then immediately permitted to inoculate healthy subjects. On August 27 Carroll was bitten by an insect that had fed, some twelve days earlier, on a patient in the second day of the disease: "In writing to Dr. Reed on the night after the incident I remarked jokingly that if there were anything in the mosquito theory I should have a good dose; and so it happened."[17] On August 31 a second positive case turned up: Dean, a soldier who had not left the camp for two months, was bitten by the same mosquito that had bitten Carrol and by another insect that had fed twelve days earlier on a patient in the second day of illness who later died. In September tragedy struck: Lazear was accidentally bitten on the thirteenth, fell ill on the eighteenth, and died on the twenty-fifth. In light of these preliminary results, it appeared that the mosquito was infectious if it fed on a disease victim early in the illness and if the mosquito itself had been previously infected. Early in October Reed returned to Havana. In feverish haste he drafted his "Preliminary Note," the conclusion of which is well known: "The mosquito serves as the intermediate host for the parasite of yellow fever."[18]

These are the facts. It remains to give them voice. What meaning might they possess other than that which is usually assigned to them? My interpretation at least offers the advantage of respecting chronology and logic. Consider, first, the Americans' sudden conversion to the mosquito theory. Little notice has been taken of the fact that the "Preliminary Note" was composed in haste and with the aid of two other documents. The first of these was the article by the English visitors, which Reed obtained from Finlay, as evidenced by a letter from Reed to Finlay dated October 7, 1990: "I have taken the liberty of sending my driver for the copy of the *British Medical Journal* containing Durham's and Myers's note."[19] The second was the famous "Pocket Notebook" in which Lazear recorded the details of his experiments, as is shown by a letter from Reed to Carroll dated September 26, 1900: "I got the general to cable yesterday about securing Lazear's notes which he wrote that he had taken in each case bitten by mosquitoes.—Examine them carefully and keep them all."[20] From the English article Reed took the sequence of ideas that led to Finlay's theory: Carter's observation, the confirming case of the two nurses, and Ross's work. From Lazear's notes Reed took the details of the experimental inoculations. In other words, Reed made the decision (Durham and Myers) coincide with the demonstration (Lazear). Out of this came the "Preliminary Note."

THE THEORY VERIFIED

At this point someone may object that just because Reed used the English article it does not follow that he was indebted to the English for his decision. After all, Durham and Myers acknowledged that the observation of the two nurses came from Reed, and Reed acknowledged that he had been pleased to report Carter's observations to the English. In the concrete situation as it existed in Cuba that summer, it was Carter's discovery that led to the mosquito theory. If we could find out who made that discovery, we would know who gave what to whom. Was it the American commission? Whoever was capable of the greater feat (reasoning from Carter's observation to the mosquito theory) was capable of the lesser (relating the case of the two nurses to Carter's observation). As it happens, Reed let slip the information needed to settle the matter: "This observation [of the two nurses] was recorded at the time for what it was worth. Later it was found to harmonize with the observation which had been made by Surg. Henry R. Carter."[21] *Later*—that is, in July. Why did the Americans notice in July a connection that had escaped their notice in June? Because the English had meanwhile brought it to their attention, along with everything else. One final detail is worth mentioning. It was Carter, not Reed, who passes his observations on to the English. And the conversation, to judge by what Durham and Myers have to say, could not have been more cordial.

By pointing out a new avenue of research, the English cleared the way for the Americans. Fear of being overtaken by the English explains the rest. Finlay, who was visited by Durham and Myers on July 25 and by the Americans shortly thereafter, was not fooled: "I believe that the haste with which this commission initially decided to verify my mosquito theory was dictated by their fear of being overtaken by the English commission from Liverpool."[22] Haste—Finlay used the word advisedly. Haste led to costly and tragic errors, and in Washington Reed suffered anxiety and disappointment, as is evident from the letter he sent Carroll on the eve of Lazear's death: "I think I have experienced more mental distress during this month . . . than ever before. . . . I am only regretting that two such valuable lives have been put in jeopardy, under circumstances in which the results would not be above criticism."[23]

In the heat of the moment Reed had his reasons for criticizing the very first experiments. The fact remains that they were carried out in his absence. Responsibility for the project rested entirely with Lazear, who of course had worked with Grassi in Rome. He was, moreover, one of the first researchers in the United States to confirm Ross's work on malaria. In June 1900 his paper on the "Pathology of Malaria Fevers, Structure of the Parasite, and Changes in Tissue" was read to the congress of the American Medical Association. It was Lazear who had worked with Thayer at Johns Hopkins. Thayer, moreover, was well aware of his role: "The part which he [Lazear] played in that work was essential and important. . . . He was one who made his own plans and worked out his own problems."[24] When Reed left Havana in early August, he left a work routine. Lazear, who clearly recognized the significance and scope of the work, took charge. He had already filled his pocket notebook with observations when Reed put to Carroll a question disarming in its naïveté: "Did the *Mosquito* DO IT?"[25]

American historians wished to celebrate the triumph of American medicine. From their standpoint the guidance offered by the English commission, the series of substandard experiments, and the absence of Reed were embarrassing. In the finest tradition of military history, they therefore idealized reality. Those who did not simply omit the contact between the British and American commissions described it as an episode of honorable battle on the fields of science. The tribute the vanquished paid to the victors seemed to preclude the possibility that the English commission, bested in battle, had played any role whatsoever: in 1902 the English published a report in which they "expressed warm thanks to the members of the Reed commission and credited that commission with many findings."[26] The early experiments then took on exemplary meaning. Carroll's illness and Lazear's death became marks of devotion and heroic sacrifice: "The heroism displayed in the investigation of the etiology of yellow fever is not surpassed in the annals of scientific research."[27] Finally, Reed's absence proved the easiest of the three difficulties to overcome. Lazear's role was reduced to that of a subordinate carrying out the orders of his commanding officer. Reed was thus credited with achievements that were Lazear's: "He [Reed] was the originating, directing, and controlling mind in this work, and the others were assistants only."[28]

THE END OF THE TALE

Lazear's experiments, as we have seen, were not beyond criticism. Unlike Finlay, he had used mosquitoes born from larvae in captivity. He failed, however, to exercise strict control over his experimental subjects, and other sources of infection could not be ruled out. Carroll, for instance, visited the autopsy room prior to being inoculated and might have contracted the disease there. Hence the first commentaries were skeptical. The *British Medical Journal* summed up the attitude of the medical world: "Unfortunately, the mode in which the experiments were conducted detracts much from their value. They are really by no means conclusive. The experimenters themselves are of this opinion. At most they are suggestive."[29] Further work remained to be done. Reed would seize the opportunity. Having distinguished himself by his absence, he would soon thrust himself into the limelight by producing a striking *confirmation* of Lazear's work.

THE THEORY VERIFIED

There is no denying the fact. The experiments that Reed performed after his return from Washington would relegate the dubious early results, as well as Reed's absence during August, to oblivion. Only the able experimenter and his striking conclusions would be remembered: "The demonstration of the Army commission is one of the most brilliant and conclusive in the history of science."[30] Reed was able to control all the variables. From now on the commission would experiment on persons of legal age aware of the risks and willing to accept them. The military government assumed responsibility for the project. Reed selected a locale supposedly free of the epidemic: "Camp Lazear," located near the city of Quemados is a spot that was sunny, adequately drained, and well ventilated. There the movements of the volunteers were closely supervised. They were subjected to quarantine before being exposed to the mosquitoes. In order to make sure that they were not ill with yellow fever in the incubation period, they were placed under medical surveillance. By November 20 everything was in readiness.

A preliminary series of experiments confirmed that the mosquito was the intermediate host of the parasite. First, six volunteers were bitten by insects that had been allowed to feed, more than twelve days earlier, on patients in the early stages of the disease (prior to the third day). Reed observed five clear cases of yellow fever. After passing through the mosquito's stomach, the parasite entered the salivary glands: "An interval of about twelve days or more after contamination appears to be necessary before the mosquito is capable of conveying the infection." Next, a large room was divided into two smaller rooms separated by a metal screen. Fifteen infected mosquitoes were set free in one of the rooms. A volunteer entered the room on three separate occasions and was bitten several times. Shortly thereafter he fell ill. Two other volunteers were placed in the adjacent room as controls. They remained for eighteen

THE THEORY VERIFIED

days without contracting the disease. The same experiment was repeated with other volunteers. Six of seven cases proved positive: "A house may be said to be infected with yellow fever only when there are present within its walls contaminated mosquitoes capable of conveying the parasite of this disease."

A second series of experiments refuted the old theory of *fomites*. With this test of an old theory Reed undeniably won the admiration of contemporaries and historians. It is not clear, however, exactly what impressed them the most. Was it the negative conclusion? "Yellow fever is not conveyed by *fomites*, and hence disinfection of articles of clothing, bedding, or merchandise supposedly contaminated by contact with those sick with this disease is unnecessary."[31] Experts had known this for a long time. Was it his experimental procedure? It might seem so, for that procedure was unprecedented: "Three young Americans slept for twenty consecutive nights in a room garnished with articles soiled with black vomit, bloody fecal discharges, and urine from fatal and other cases of yellow fever. . . . Not the slightest indisposition followed close and intimate contact with this repulsive material in any case."[32] Did not the proof of the mosquito's role contain a refutation of the theory of *fomites*? Reed went too far—and in so doing conducted himself, if I may say so, like the ragpicker of tropical medicine.

Malaria continued to serve as a guide. Blood taken from yellow fever victims in the early stages of the disease was injected intravenously. In four experimental inoculations with quantities ranging from 0.5 to 1.5 cubic centimeters of blood, the commission noted three positive cases. These experiments showed that the parasite was present in the blood in the early stages of the illness and that it could be transmitted "like the malarial parasite, either by means of the bite of the mosquito or by the injection of blood from the general circulation."[33]

THE THEORY VERIFIED

This result revived interest in the etiological question. Welch is credited with having drawn the commission's attention to the idea that the agent of yellow fever might be a virus. But Novy had already made such a suggestion. In any case, blood from a victim was placed in a sterile test tube and the serum separated out. Injection into a healthy subject resulted in an experimental case of the disease. Then the serum was heated to fifty-five degrees for ten minutes; this time inoculation had no effect. Hence the virulence was not due to a toxalbumin. Finally, the yellow fever virus passed through Berkefeld's filter and Chamberland's filter. Injection of the filtrate of diluted serum resulted in a characteristic attack: "Yellow fever, like the foot-and-mouth disease of cattle, is caused by a microorganism so minute in size that it might be designated as ultramicroscopic."[34]

The work of the Americans seemed to mark a turning point in the history of research on yellow fever. An era had ended, apparently, and a new day had dawned. As Soper so unfortunately expressed it, the "Dark Age" was over. The Reed commission's results became the standard of comparison for all subsequent work. Numerous confirmations were soon forthcoming. In Cuba Guiteras and Finlay won renown by producing three fatal cases. In Brazil Ribas and Lutz confirmed the American findings, as did the better-known Marchoux Commission (Simond and Salimbéni) sent by the Institut Pasteur. In Mexico an American commission sent by the Public Health Service and Marine Hospital produced further confirmation. And the consecration of practice was quickly bestowed on the theory. Now that the intermediate host was known, yellow fever could be stamped out. Gorgas eliminated the scourge from the island within six months. It sufficed to break the cycle by destroying mosquitoes and their larvae.

THE THEORY VERIFIED

Some historians have looked upon the work of the American Yellow Fever Commission as an outgrowth of Finlay's memoir of 1881. In their view the commission resurrected a hypothesis that had lain in oblivion for twenty years and proved that it was correct. In reviving Finlay's hypothesis, however, the Americans, it is alleged, had to overcome several major hurdles in the form of opposition to the Cuban physician's theory stemming from opportunism in the field of preventive hygiene, scientific pragmatism, and ignorance. These fictitious hurdles were created purposely to justify claims of a continuity that did not exist.

To write a linear narrative one would first have to show that Finlay and Reed were working on the same problem. It will emerge in due course that Finlay's work was separated from that of the American commission by an unbridgeable divide. Ross's work marked a watershed. Finlay believed that the mosquito served as a carrier. By contrast, the American commission believed that the insect played an essential part in the life cycle of the parasite. At the time, therefore, the Cuban physician's hypothesis met with justifiable resistance.

It is nevertheless true that Finlay's work, along with Manson's, contributed to the formulation of questions about disease vectors. Manson and Finlay showed that the origins of infection might depend on the interaction among living things. Thus they helped medical thought to overcome a block that had prevented scientists from conceptualizing the role of arthropods in the transmission of disease between humans.

ILLUSORY CONTINUITY

Historians were quick to see a connection between Finlay's discovery and the demonstration carried out by the American commission. Finlay's work appeared to anticipate and imply the ultimate result; the Cuban physician seemed to have supplied the theory and even the methods that led to the great achievement. The work of the Americans was seen as a direct outgrowth of Finlay's theory: "It was in 1881 that Finlay first announced the hypothesis that a mosquito, *Culex mosquito* (*Stegomyia fasciata*) was the vector of the infection. This hypothesis was to receive striking confirmation at the hands of Reed, Carroll, Agramonte, and Lazear."[1]

This view of the matter is clearly based on a methodological bias in favor of historical continuity. According to this view, the pattern of discovery is orderly and sequential; time's only role is to measure the lapse between one discovery and another. For the matter at hand the chief consequence of this methodological presupposition was that it confronted historians with a need to account for the two decades that elapsed between the statement of the hypothesis and its exemplary confirmation. Compared with this long period of latency, the speed of confirmation was truly impressive. It took no more than two months to provide proof of Finlay's hypothesis. Such rapid verification suggests that there was nothing terribly difficult about the methods used. If triumph was indeed so effortless, then the cause of the delay can only lie in the scientific community's indifference to Finlay's theory. Historians forced by the logic of their method to this conclusion must then account for such peculiar disdain on the part of men of science. Some argue that the American authorities were not particularly concerned about yellow fever. Others maintain the vogue for bacteriology eclipsed the Cuban doctor's theory. Still others insist that Finlay's hypothesis was so ingenious that it was misunderstood.

These explanations have been repeated time and time again. It has been argued, for example, the eliminating yellow fever was not a priority of American health institutions: "It is clear that, in the broadest sense, factors outside science, such as political indifference to yellow fever . . . dictated the delay in solving the yellow fever problem before 1900."[2] Yellow fever was seen as a regional and not particularly dangerous disease. It could be combated by simple disinfective measures and avoidance. Another common argument is that physicians were bent exclusively on discovering the yellow fever germ: "It seems inconceivable today [that] Finlay's discovery, announced in 1881, was for two decades rudely ignored. . . . A very important factor was the great enthusiasm of the time for germ theory. . . . The *development* of germ theory . . . blinded his illustrious colleagues to the idea."[3] Finally, it is claimed that Finlay's theory was not understood: "His concept was so ingenious, so far beyond what was known at the time, that it was not accepted because no one understood it."[4] Out of this came the myth of the lonely genius whose brilliance was rivaled only by the indifference of the scientific community: "The solemn hour had yet to strike, when the glass of human incredulity would be broken and resistance to progress overcome."[5]

At length the moment came for eyes to open. The first factor that aroused the attention of the American authorities was the military occupation of Cuba. A dramatic increase in cases of yellow fever was observed despite an exemplary sanitation campaign. The pressure of interests was therefore decisive in identifying the problem, seeking a solution, and putting it to practical application. American attitudes changed in three ways: yellow fever was now perceived as a danger, the research budget was increased, and the occupation made possible a systematic application of the results: "In the new political and economic conditions of 1900, the Reed Board confirmed the mosquito theory in two months."[6] The second factor was the failure of bacteriological research: "The idea came to Dr. Reed after the board had demonstrated that the claim of Sanarelli . . . was without foundation."[7] The third factor that led physicians to reconsider Finlay's theory was knowledge of the transmission of disease by insects. It took twenty years for other scientists to achieve the level of scientific knowledge that Finlay had attained in 1880: "Conditions had changed, since Finlay had started from zero . . . whereas the physicians of the American board had long been familiar with the work of Manson, Loos, Ross, Laveran, and Rossi."[8]

These reconstructions are historically false. It is simply inaccurate to say that during the last two decades of the nineteenth century the epidemiological question did not arouse the slightest interest. The work of the Chaillé Commission (1879), Sternberg (1890), and Wasdin and Geddings (1897) proves the contrary. Furthermore, it is quite naïve to think that, had the will to solve the problem only existed, the solution would have presented itself immediately. Political practice did not shape or alter the theoretical structure of tropical pathology. Nor is it correct to say that it was necessary to disprove the alleged role of the Sanarelli bacillus before Finlay's theory could be rediscovered. If curiosity in Finlay's work stemmed from the failure of bacteriological research, it should have been awakened much sooner—in 1890, say, when Sternberg disposed of all the pseudo-discoveries of alleged germs. What is more, the bacteriological possibility was never abandoned: the commission continued its bacteriological research even after it undertook to confirm the mosquito theory. Finally, it cannot be maintained that Finlay was the Semmelweis of tropical medicine. A glance at the medical literature of the time shows that his work was known and discussed abroad. He was not the victim of a conspiracy of silence or of an intellectual incapacity to comprehend.

THE WATERSHED

Finlay published his memoir in 1881. From that time on his theory was undeniably common knowledge in the medical community. Reed did not rediscover Finlay after a long period of stubborn indifference. There was no latency period between Finlay's hypothesis and its *utilization* by Reed. The error, and the reason why it has been made so often, lies here: in the belief that Reed *used* Finlay's theory. That may seem strange to us today, now that the concept of vector, which incorporates both the idea of a host and that of a vehicle, is so firmly established. But those two ideas differ in age and in meaning. The notion of an agent of transmission involves a mechanical carrier. Finlay's research had nothing to do with the problem of the life cycle of the yellow fever germ. The question never even came up; the germ had not been identified.

Ross changed the whole picture. His discovery of the hematozoon's life cycle *revealed* malaria's mode of infection. Twenty years was the time it took to solve this problem. The discovery, as we saw earlier, proceeded from Laveran to Ross by way of Manson. Then and only then could the mosquito shed the role of passive conveyor to acquire that of intermediate host. Ross, always quick to defend, indeed to expand, his little empire, was under no illusion as to the significance of the American commission's work. His esteem for that work no doubt owed much to the fact that it bore his imprint: "They determined to investigate by direct experiment the carrying capacities of mosquitoes—on the basis, not of Finlay's ideas, but of my work on malaria, according to which the virus must develop for some days within the carrying mosquito."[9]

Strictly speaking, neither Finlay nor anyone else could have formulated the hypothesis of the mosquito as intermediate host. Besides the fact that the germ was unknown, there was no basis for drawing an analogy: something was missing. Before Ross's discovery, nothing was known about the hematozoon's relation to its host. Very little was known about the filaria's relation to the mosquito either: that question was not settled until after the mystery of malaria had been unraveled. Before 1898 there existed no research object to which the yellow fever germ might have been compared. Before that date there was nothing in tropical medicine capable of *reorienting* research in the direction of the *Culex* mosquito as intermediate host. Ross's work therefore marked a watershed. If Reed succeeded where Finlay failed, it was because he took from Ross the conceptual tools necessary to produce an experimental infection.

True, the American commission worked with material supplied by Finlay: the correct species of mosquito. But while it preserved the referent, it altered the content of the reference by substituting the notion of host for that of carrier. In order to claim that the American commission confirmed Finlay's hypothesis, this substitution must be neglected. One is then led, along with Dominguez Roldan, to express astonishment: "I honestly cannot see how Reed's American commission could have discovered the same facts . . . and yet dared to proclaim that it had invented them."[10] By contrast, remarking the substitution dispels many misconceptions. If we admit that Reed did not verify Finlay's hypothesis, we can see what Dominguez Roldan could not: that proving that the mosquito was an intermediate host was not a matter of discovering the same facts as Finlay. After the fact it was of course possible to say that Finlay had captured the host without recognizing it. But that would only disguise, not eliminate, the difference: Finlay missed what Reed grasped.

It may nevertheless be objected that the American commission did indeed confirm Finlay's hypothesis: the notion of host subsumes that of carrier, and Finlay never ceased to argue that the mosquito serves as a vehicle. Once the mosquito theory was verified, it was tempting to see Finlay's hypothesis as a part of that theory. But to do so requires prying apart two intimately related notions. To the extent that the notion of vehicle *derives* from that of host, it must not be confused with the idea of a carrier. The expulsion of parasites by the salivary glands is to the regurgitation of parasites as a biological process is to a mechanical operation. Nor can it even be argued that the American commission *rediscovered*, by another route, what Finlay had already found out. The idea of a vehicle figures as an *implication* of what it is supposed to prove: that the mosquito functions as an intermediate host. Furthermore, the question that Finlay answered with difficulty—and brio— the American commission never even raised. The solution would emerge automatically once the life cycle of the yellow fever germ was established: to discover the cycle was to indicate the vehicle.

Where Reed seemed to be adding the notion of host to that of vehicle, he had already covered the same ground as Finlay, but without following in Finlay's footsteps. The Cuban doctor's theory did not in broad outline prefigure that of Reed. The work of the American commission was not the culmination of a slow process of refinement; the distance from Finlay to Reed was not traversed by amassing data or rectifying a few principles. The recasting was complete. The American commission did not base its research on the idea of the mosquito as vehicle. Its initial hypothesis was this: the specific agent belongs to the group or protozoa.

Those Cuban historians who have attempted to reduce the American commission's role to one of confirmation were misguided. Inevitably they missed the key condition that made the whole enterprise possible: the substitution of Ross's theory for Finlay's. Inevitably, too, they saw the course of the investigation as being determined by decisions made by individuals other than Finlay. This accounts for their strange but insistent reproach to the Americans: "Sternberg's error was grave. On several occasions he had the opportunity to study and prove Finlay's theory and to advance the progress of medicine by nineteen years."[11] Things were not that simple. It must be conceded that the Cuban physician did not possess what the American commission found in Ross. What Finlay possessed was simply this: the idea that the mosquito is a carrier of yellow fever and inoculates victims with germs accumulated in its proboscis.

It follows that the skepticism with which Finlay's hypothesis was greeted was justified. The usual verdict must therefore be reversed. The extreme skepticism of Finlay's contemporaries was a sign not of blindness but of astonishing lucidity. No other reaction was warranted. Finlay had avoided the central problems of classical epidemiology but at a relatively high price. He had had to revive the old theory of vaccination, which meant accepting the inevitable inconsistency associated with this model and the failure inherent in its application. Inconsistency, in that likening the mosquito to a syringe bestowed a random character on the transmission of the disease. And failure, in that the analogy with Jenner suggested an experimental protocol that confused hypothetical vaccination with experimental demonstration. Finlay's contemporaries, not being fools, directed their criticisms at these two weak points in his exposition: the randomness of transmission and the absence of proof.

Consider first the discussion of the mode of propagation. Vaccination was the artificial form of a natural process, a technique that, by mimicking nature, directed and made use of its powers. This was the model that guided Finlay when he placed a contaminated mosquito on the arm of a healthy individual. In so doing, Finlay did not for one moment doubt that he was imitating what occurs *spontaneously* in nature, which either bestows immunity (through one bite) or transmits the disease (through more than one bite). By assuming a similarity, indeed an identity, between his own technique for handling contaminated mosquitoes and what takes place in nature, however, Finlay laid himself open to criticism. And here his theory came up against its unsurpassable limitations.

In one respect, Finlay's technique was based on the vaccination model, since the mosquito's proboscis was likened to "a new type of inoculating needle."[12] But Finlay projected onto nature an appearance of finality that fooled no one: an inoculating agent lacking the *equivalent* of the intention that guides the vaccinator's hand is an exceptional, not to say accidental, procedure. Béranger-Féraud made the point: "Not for a moment would anyone believe that that is the ordinary, normal means of transmitting the disease."[13] In another respect, Finlay's epidemiological research came under the head of medical geography, since he accumulated observations intended to show that the distribution of the disease coincided with that of the mosquito. This coincidence argued in favor of his hypothesis, but it did not prove the existence of contamination by way of inoculation. Once again Béranger-Féraud summed up the objection with a play on the meaning of "coincidence": these observations, he said, "are in my opinion nothing but pure coincidence."[14]

Finlay, moreover, was well aware of the contradictions in his theory. If the proboscis of the *Culex* mosquito acted as a syringe, not only could it carry the germs of the other diseases of the blood, but any species of mosquito ought to be able to transmit yellow fever. In order to argue that the disease is transmitted solely by the *Culex* mosquito, it had to be shown that that species was somehow "fixated" on the yellow fever germ. Finlay conceived a process that was at once elective and selective. The salivary secretions of the *Culex* mosquito must exhibit two qualities: they must be "germinative" for yellow fever germs and "bactericidal" for all other germs. When Finlay placed the heads of infected *Culex* mosquitoes in sterilized test tubes, he found his specific micrococcus to be present and all others absent. By contrast, no micrococcus at all was present in cultures prepared from other species of mosquitoes that had been allowed to feed on yellow fever victims.[15] Clearly Finlay interpreted the experiment in a way that raised more questions than it answered.

Let us turn now to the discussion of inoculation. Although Finlay claimed to have produced attenuated forms of the disease, he knew full well that decisive proof in the form of an unmistakable case eluded him. At the time he evinced an interest in the very dubious experiments of Carmona y Valle, which suggested that yellow fever was transmissible by inoculation. Later, however, Finlay would transform his inability to transmit the disease in a clearcut way into a moral imperative: he claimed that his objective had been to avoid rather than provoke a serious attack. Contemporaries were not convinced that inoculation worked either as a technique of vaccination or as a mode of transmission. Sternberg saw no reason why the virus taken up along with the blood should lodge in the insect's mouth apparatus: "It [the blood] enters the insect's stomach, and whatever remains after its meal has been digested is passed *per anum.*" If, moreover, the virus was transmitted through the mosquito's bite, the attempts at artificial contamination should have succeeded. But the results of experiments performed at Veracruz were negative: "Some experimental evidence . . . indicates that the blood of an individual sick with yellow fever may be injected beneath the skin of an unacclimated person without producing any noticeable effect."[16]

Even if these experiments had succeeded, moreover, Finlay's theory would not have been confirmed. The American commission showed, as we saw earlier, that yellow fever, like malaria, could be transmitted through inoculation of blood drawn from a victim. In 1901 Finlay felt that if it had been known that the disease could be transmitted in a few drops of blood, not only would his theory have been accepted, but someone would have calculated the number of bites needed to obtain an unambiguous reaction. Oddly enough, Finlay as yet failed to see that the medical community would then have confronted a problem similar to that which had arisen in the study of malaria: the question of the natural mode of propagation of the disease would have remained unanswered. This alone suffices to show, yet again, how far Finlay was from conceptualizing the illness in terms of a parasite with alternating hosts. Finlay would ultimately acknowledge that the mosquito served as intermediate host yet still not abandon his hypothesis, because he still clung to the belief that successful inoculation of the disease by means of a hypodermic needle validated his old idea. To Finlay, the laboratory experiment was still just a model for another natural procedure. In 1901 he argued that the Americans were wrong "to deny the possibility of direct inoculation through the needle nose of the mosquito, since they themselves have carried out the operation with the blood of yellow fever victims."[17]

The attacks on Finlay by his contemporaries were thus not without foundation. Still, they were not entirely justified. His work, like Manson's had resulted in an epistemological transformation that was historically and concretely a prerequisite to the development of a new perception of tropical disease. Ross remarked that prior to Finlay's memorable work "no one dreamt the mosquito could play two roles . . . at once taking the parasite from the patient and inoculating it into healthy individuals."[18] And Reed wrote: "Six months ago, when we landed on this island, all was an unfathomable mystery, but today the curtain has been drawn, this mode of propagation is established, and we know that minuscule mosquito is no more."[19] The sudden amnesia that afflicted two of Finlay's most illustrious successors bears indirect witness to the value of his work. Not only had Finlay "dreamt" that the mosquito might fill both roles, but also Reed had received from him an account of that dream, which remained in need of interpretation, along with the correct species to look at.

We have seen how much Ross profited from his collaboration with Manson and Reed from the support of Finlay. We also know how Manson and Finlay were compensated for their trouble: with words of denigration. Ross (at the end of his life) said that Manson had not known how to interpret the phenomenon of exflagellation and had understood nothing about the parasite's mode of entry. And Reed repeatedly stated that Finlay had succeeded only in discrediting his own hypothesis. To bolster their slender claims to originality the scientists of the moment attempted to discredit those who had preceded them. Those predecessors cast a shadow over them—not unjustifiably. What mattered in the end was not "Manson's error" or "Finlay's error" but the perception that showed they had been right in spite of themselves.

THE ANALYSIS OF DIFFERENCES

The purpose of deciphering epistemological transformations is not to rehabilitate reputations, however. At issue is the recognition that arthropods have an important place in medical science, a subject that calls for a few additional remarks. In the case of filariasis, we can say that Manson was stumped by a problem of epidemiology; he was never able to use his zoological knowledge to demonstrate the transmission of the disease in a satisfactory way. He believed that the microfilariae required an intermediate host in order to complete their metamorphosis, not in order to be reintroduced into the human body. In the case of yellow fever, we can say that Finlay was stumped by the problem of the life cycle; not even in a hypothetical way was he ever able to bring his epidemiological knowledge to bear on the problem of parasites with alternating hosts. Finlay believed that the germ required an agent in order to be transported from one individual to another, not in order to complete its own life cycle.

By contrast, Ross and Reed said that the mosquito is the intermediate host and therefore the medium. The preeminence of the zoological view led Ross to discover the life cycle of a protozoon and Reed to adopt a new working hypothesis. Just as the discovery of the hematozoon's life cycle was the key to understanding malaria's mode of infection, so, too, did Reed's new hypothesis prove to be the key to understanding that of yellow fever. The difference between Ross and Manson and between Reed and Finlay is immediately apparent. Ross differed from Manson because he completed the account of a life cycle that Manson had never fully comprehended. And in completing the life cycle Ross also discovered the pathway by which the parasite entered the body, something that had always eluded Manson. Reed differed from Finlay because he based his work on a hypothesis that Finlay was never able to formulate. By applying the hypothesis, Reed succeeded where Finlay had invariably failed: he produced an experimental case of yellow fever.

Ross seemed to revert to themes already explored by such contemporaries as Küchenmeister, Steenstrup, and von Siebold. Hence his work appears to belong to the classical tradition of parasitology. According to this view, his achievement was simply to have been the first to incorporate a protozoon into the general theory of parasites with alternating hosts. Reed also seemed to revert to themes previously explored by Budd, Pettenkofer, and Bemiss. His work therefore appears to belong to the classical tradition of epidemiology. According to this view, his achievement was simply to have given a name to the mysterious nidus: a living insect and not an inanimate environment. Combining these two views gives rise to an appealing version of history. In the 1880s physicians compared the yellow fever germ to the germs of cholera and typhoid fever and concluded that it must evolve in the environment in a manner similar to the helminths. By the end of the century, after Ross had clarified the hematozoon's cycle, Reed presumably had only to make the assumption that the yellow fever germ behaves like a protozoon to discover its intermediate host.

If one accepts this view, it is possible to trace three avenues of progress in tropical medicine: one from Leuckart to Ross, a second from Bemiss to Reed, and a third from Ross to the American commission: "A new and fertile field for research made possible by the brilliant work of Ross . . . has recently been further extended by four American investigators."[20] The problem with this version of history is not only that Manson and Finlay are entirely eliminated but also that the descriptive account is misleading. Instead of seeing the process as one of repetition and extension, we must look at the fine detail in order to discover what was actually a process of transformation.

The themes of classical parasitology and traditional

epidemiology do not coincide with the themes we find in the work of Ross and Reed. Knowledge had to be recast before the role of vectors as understood at the end of the nineteenth century could be identified. And that recasting of knowledge depended on the work of Manson and Finlay.

Classical parasitology was *the science of organisms whose biotope was alive:* the study of the single- or multiple-host life cycles of parasites. Microbiology was *the science of disease agents:* the study of the causes of organic alterations and infections. Parasitology and microbiology therefore did not have the same object of study, as the work of Leuckart on the one hand and Pasteur on the other makes clear. A science centered on the natural history of parasites *as living organisms* was opposed to a science centered on the natural history of microbes *as aggressors.* In the first case the living hosts constituted the links of the chain; in the second, the four postulates of Koch. Parasitology identifies the strategies by which living things carry out their functions. By contrast, microbiology is concerned with the elaboration of a strategy to combat that which poses a threat to man.

When the two disciplines are contrasted with respect to their relation to epidemiology, however, it becomes clear that they intersect, and indeed fit together so well that the definition of classical epidemiology applies equally well to both. Epidemiology was *the science of vehicles of parasites and microbes as disease agents.* Out of this arose a fundamental distinction between contagious diseases, in which man serves as the carrier of the pathogenic agent, and indirectly contagious disease, in which the environment serves as a medium between transmitter and receiver. In one respect epidemiology was Hippocratic, in another, Newtonian. It was Hippocratic because of its anthropocentric component: man was seen as the point of diffusion and convergence of disease agents. It was impossible to see man as the agent's host and thus to develop an epidemiology conceived in terms of the microbe or parasite. It was Newtonian because of its mechanical component: the vehicle of the pathogenic agent was seen as a material and passive medium. It was impossible to see the medium as a vital and active intermediary and thus to develop an epidemiology conceived in terms of the vehicle.

The solidity and autonomy of classical epidemiology's conceptual framework impeded the development of a science of vectors. It could not accommodate the two points of view that had to be taken into account in order to envision an arthropod as host or vehicle. Parasites and microbes were always carried by inanimate objects: tainted food, contaminated flesh, or drinking water. It it were necessary to sum up the epidemiology of parasitic and infectious diseases in a phrase, perhaps the best description would be that "illness enters and exits man as if through a door."[21]

Because the transformation in question occurred partly within the field of parasitology, we must consider the relation between that discipline and epidemiology. The two disciplines offered precisely overlapping yet independent explanations. Parasitology studied the parasite as a living thing, while epidemiology spoke from the human standpoint about the vehicle of the parasite as pathogenic agent. Because the two explanations were independent, the epidemiological story could not be told exclusively in terms of the parasite. But because the two explanations overlapped, parasitology was obliged to work within definite limits. For one thing, it was inevitable that the entry and exit pathways of the parasite would be limited to the alimentary canal. Hence the system was closed in two senses. First, the intermediate host was necessarily assigned the status of human prey: being passive, it has to be swallowed along with the parasite it carried. Second, the intermediate host necessarily had to take up the parasite along with its food and drink from the environment into which it had been expelled by man. The life cycle of the parasite was thus always contained within the food cycle.

The science of vectors is nevertheless rooted within this discursive structure. In order for roles to be reassigned, a cycle different from that defined by the helminths had to be found. In other words, the mechanism of classical parasitology had to be overturned. Ross, by denying both that the epidemiological and parasitological accounts had to overlap and that they were independent, expanded the parasitological view beyond the epidemiological, which thereby became subordinate. Ross deduced the epidemiology of malaria from the life cycle of a protozoon. By showing that the first thing to investigate was the life cycle of the parasite, he ultimately conflated the concepts of vehicle and intermediate host. After Ross the zoological and epidemiological threads are woven into a single strand.

This transformation, which culminates in Ross's work, was largely adumbrated by the work of Manson and Finlay. Manson anticipated Ross by attempting to show that an epidemiological account could be deduced from a parasitological one. The search for the filaria's host introduced a new prospect: epidemiology seen from the standpoint of the nematode. From this came the notion of the mosquito as intermediate host along with a breach in the edifice of classical parasitology. Thus there was a diversification of the procedure of transmission or, perhaps more accurately, the emergence of a new process: in addition to taking the parasite from the environment into which it had been expelled by man, the intermediate host could also take it directly from the blood. The introduction of the hematophagous insect soon enabled Finlay to look at the epidemiology of yellow fever from a new angle. His study proposed the mosquito as agent of transmission. The search for the carrier opened up a second prospect: epidemiology seen from the standpoint of the vehicle. This time the breach was in the edifice of traditional epidemiology: the vehicle was not simply a material, passive medium but could also be actively involved in taking up and inoculating the pathogenic agent.

To the helminth cycle involving passive intermediate hosts Manson added the concept of an active intermediate host. To the transmission of indirectly transmissible diseases by inanimate carriers Finlay added the concept of an animate vehicle, a biting invertebrate. To measure the extent of the transformation one must refer back to the lapidary formula of Canguilhem cited earlier: "Illness enters and exits man as if through a door." But here the door is closed. Illness enters and exits like a burglar with a skeleton key. Such was the transcutaneous pathway discovered by Manson and Finlay.

Schematically we may say that the problem confronting both Manson and Finlay was the same: to prove that the filaria or yellow fever germ passes from the circulatory system of one individual to that of another. While contemporary researchers stuck to the usual pathways, these two men first noted impossibilities. Both began their conceptual transformations by ruling out classical solutions. A transformation was implicit in Manson's question: given the size, structure, and localization of the microfilariae in the closed circulatory system, the problem was to find a host capable of facilitating their exit and furthering their life cycle. Finlay posed a similar question when he assumed that the inner walls of the vessels constituted the germ's habitat: given the localization of the virus in the vascular endothelium, the problem was to find an agent capable of removing it and injecting it into the corresponding tissue of a healthy individual.

The two problems gave rise to two different conceptual frameworks. Manson looked to the mosquito as intermediate host. Finlay rejected that answer, since the germ was unknown, and looked to the mosquito instead as agent of transmission. Both the biological and the mechanical solution served to resolve what was in effect a single problem resulting from a single transformation: by what pathway the disease was transmitted, given that none of the usual pathways worked. The biological solution called for observation of the larva's metamorphosis inside the mosquito's body, while the mechanical solution led to observation of the insect's habits. The first approach enriched parasitolology by adding the concept of a bloodsucking insect as intermediate host. The second enriched epidemiology by enabling it to adopt the point of view of medical entomology.

This transformation required the application of two complementary principles of biological epistemology. First, the requirements of logic had to be respected. Courtès points out the good—and neglected—aspect of this principle: "Namely, that logic does not so much prescribe as prohibit and contradict."[22] Second, the requirements of life also had to be respected. Canguilhem points out the good—and neglected—aspect of this principle: "It is the living thing that, through its structure and functions, provides the key to its own decipherment."[23]

The requirements of logic led Manson and Finlay to discover the weak points in the classical conceptions. Manson ruled out the possibility that pathological secretions and discharges provided the parasites with their exit pathway. Similarly, Finlay denied that such effluvia were the vehicle and hence that the air or water played a part in the process. As for the requirements of biology, it became clear that looking at the question from the standpoint of the animal was incompatible with looking at it from the standpoint of man. Manson saw the filariae as pursuing the functional goals of self-preservation and reproduction, and similarly Finlay defined these functions in the mosquito as objectives to be attained. For Manson it was a question of finding the host in which the microfilariae could effect their metamorphosis. For Finlay it was a matter of linking the insect's meal of blood to ovogenesis and the maturation of its eggs. In the work of both men we therefore find keen awareness of the logical impossibilities associated with "a kind of 'view,' in Goethe's sense, a view that never loses contact with highly empirical facts."[24]

We can now see how Manson and Finlay stood out from their contemporaries. The others never really confronted the problems raised by the filaria's life cycle and the transmission of yellow fever. For them, the problems were not so much problems as misimpressions to be corrected or gaps to be filled in with the aid of classical schemes. This avoidance of the issue accounts for resistance that stemmed from what we may call the archaeological as distinct from the conceptual level. The reaction of Leuckart, an eminent authority on parasitology, is revealing in this respect. He rejected Manson's solution and continued to believe in the usual exit pathways. In 1886 he wrote: "There would yet remain a possibility that the embryos evacuated with the urine . . . may be transported to certain small hosts, and by these means human beings may perhaps be infected more commonly than in the way pointed out by Manson."[25] Sternberg's opposition was equally significant. He rejected the solution proposed by Finlay and continued to believe in the classical pathways of transmission. In 1897 he was still comparing yellow fever to cholera and typhoid fever: "Epidemics usually extend from *foci* of infection. . . . It seems extremely probable that this occurs in the same way as in the disease mentioned . . . viz., through the excreta."[26]

By linking the filaria and the yellow fever germ to an arthropod, Manson and Finlay restructured the epistemological field. The medicine of vectors found its niche in this restructured field, although not without modifications, in particular the methodological modification introduced by Ross. The transformations carried out by

Manson and Finlay had their limitations. In order to bring the process he had disclosed back into line with classical epidemiology, Manson was forced to borrow a medium from the classical schema, namely, the drinking water. Hence the mechanism he proposed revived the old hydrological doctrine. In order to reconcile the mode of transmission he had disclosed with the fact that the etiological agent was unknown, Finlay was forced to borrow his medium from the vaccination model. Thus the mechanism he proposed revived the contagionist theme in its technological version, inoculation. Means and mechanisms were supplementary hypotheses that both men introduced to complete their systems. These borrowings were the last links between the new discoveries and classical epidemiology.

In 1898 Ross succeeded in deducing the epidemiology of a disease solely from parasitology. He was able to do this because he worked on a protozoon: an object of study that was more problematic than a nematode, because he could not simply observe the sporocysts in the mosquito's body, but less problematic than a virus, because he was able to trace the process all the way to its end in the salivary glands. Manson's seeds bore fruit before Finlay's: the introduction of the concept of mosquito as intermediate host was followed by its use by Ross, whereas Finlay's idea could come to fruition only after Ross had done his work.

In effect, the precedent of Ross enabled Finlay to function as a guide. Subsequent research would follow the path that Finlay had pointed out and work with the species he had selected. Ross thus intervened twice in the determination of yellow fever's mode of transmission. First, his work on malaria suggested the existence of a similiar process in yellow fever. In effect, it made Finlay's work credible: thanks to Ross, Finlay's lead was followed and his material accepted. Second, Ross provided the American commission of inquiry with its working hypothesis. Some historians are still partial to biographies that glorify the great geniuses of medicine. They will argue that I am giving too little credit to Finlay and even less to Reed. In response I cite this letter from a person well-versed in tropical medicine—and in matters of historical responsibility: "Nevertheless, by giving the latter [Reed] the hint and the material, you [Finlay] have done your share and it is a notable one. Many men bring the bricks to build the house, and I do not see why the men who carry the corner stone should get all the credit."[27]

The Spoils of Discovery

It was not long before Finlay claimed to have anticipated the Americans and Reed provided a version of history from which Finlay was absent. Finlay and Reed were unable to settle the dispute over their respective contributions. The reason was quite simple: they did not understand that the commission's work could be interpreted as both confirmation and refutation of the hypothesis that the mosquito is the agent of transmission. The exigencies of logic therefore led each man to choose the thesis most favorable to himself and to develop its implications. From these came fictitious reconstructions of the events.

Historians, in turn, have erected legends upon these fictitious reconstructions. Cuban historians have extolled Finlay's work and denigrated Reed's. American historians have done the opposite. Both camps employed the same methods: by adducing precursors they denied the originality of the other side and, conversely, by isolating scientific work from its context they claimed originality for their own.

Drained of all complexity and thereby impoverished, these historical legends have been put to other, nonscholarly uses. The names Reed and Finlay were soon invoked in political controversy. Following the strange fate of these two highly symbolic figures will provide us with insight into the shifting relations between Cuba and the United States.

THE FIRST INTERPRETATIONS

Havana, December 1900. Wood, the governor general, holds a banquet in Finlay's honor. Everyone is there: the Cuban elite, officers of the American army, and members of the commission of inquiry. Guiteras delivers a speech that is careful of everyone's feelings: "The glory, Dr. Finlay, is sufficiently great that you can easily share it with those who have completed your work."[1] But between the Cuban physician and Reed, hatred was mutual: fame cannot be shared. In Finlay's opinion the American commission's enterprise was inspired by base motives, and Lazear's death was hardly to its credit. In a flurry of letters Finlay and Delgado let their fantasies run wild: "Apparently the time has come when the mosquito with its little trumpet will sound my hymn of glory."[2] In Reed's opinion Finlay had met with failure. The American therefore had no compunctions about taking credit for the discovery. In a letter to his wife he wrote: "Rejoice with me, sweetheart. . . . I could cry for joy that heaven has permitted me to establish this wonderful way of propagating yellow fever."[3] It is tempting to explain both men's attitudes as the result of frustration: Finlay saw someone else complete a project he had conceived, and Reed carried out work based on an idea that was not his own. But in the end frustration must be rejected as an explanation.

The controversy soon became public. Finlay and Reed soon published articles as closely argued and persuasive as they could make them. Finlay claimed to have anticipated the commission's results and forcefully insisted on his priority. Reed, for his part, claimed full credit for the discovery and implied that he owed absolutely nothing to the

Cuban physician. Yet the facts of the matter are clear. The idea that the mosquito plays a role in the propagation of the disease was Finlay's, announced a long while before. But the verification of the theory that the mosquito is the intermediate host was provided by the American commission. What Finlay had proposed in 1881 and defended ever since is not to be confused with what the American commission stated and proved in 1990.

Furthermore, Reed and Finlay were well aware of the significance and limitations of their respective contributions. Reed knew what credit was due the Cuban physician: "To Dr. Carlos J. Finlay, of Habana, must be given, however, full credit for the theory of the propagation of yellow fever by means of the mosquito."[4] And Finlay, for his part, knew what credit was due the Americans: "I had not realized, it is true, the necessity for an incubation period in the mosquito."[5] In short, Reed knew that the Cuban physician had pointed the way, and Finlay knew that the designation of the mosquito as an intermediate host was an innovation. Why, then, did they expend so much effort claiming credit for more than they had actually done? The answer is clear: neither Reed nor Finlay understood the axiom of historical epistemology that would have justified a sharing of responsibility: namely, the American effort justified, *a posteriori*, the Cuban doctor's hypothesis but at the same time discredited it. This axiom can be taken in two ways. It can be interpreted as establishing the need for a sharing of credit, in which case there is no need for debate. Or it can be interpreted as a reason to conceal responsibilities, in which case controversy is inevitable.

If credit is to be shared, an apparent contradiction must be dispelled. The value of the Cuban's hypothesis can be recognized without ignoring its limitations. The hypothesis grew out of the merging of two lines of research; functionally it remained useful even after its content had ceased to be relevant. Finlay's hypothesis was justified in its aim (to establish a connection between the insect and the propagation of the disease) and discredited in its claim (that the germ was transported mechanically). Thus it was inevitable that the American commission would encounter Finlay's work and then set it aside. At the meeting with Finlay the right path was pointed out and the right material object was handed over. But it was from Ross that the American commission took its working hypothesis.

If credit is to be concealed, a real contradiction must be eliminated. Finlay feared that the discrediting of his hypothesis would lead to a discrediting of all his work, so that he would simply be left out of the historical record. Finlay avoided this danger by claiming total justification. He tried to show that he had stated "correct conclusions" and therefore that he had anticipated the Americans. Conversely, Reed feared that justification of Finlay's hypothesis might result in Finlay's getting all the credit, and that he, Reed, might then find himself left out of the historical record. Reed avoided this danger by attempting to discredit Finlay totally. He tried to show that Finlay had been in error and that the American commission had therefore hit upon the mosquito theory independent of Finlay's work. The dispute over priority was merely superficial.

Initially Finlay claimed full credit. The Cuban physician knew, however, that he had not provided the idea of the intermediate host. In order to get around this difficulty, Finlay claimed that he had nevertheless considered the notion and pointed out that *in 1886* he had written: "The sting often retains scores of microscopic fungi. . . . The sting of the mosquito . . . may constitute an appropriate soil for the preservation or even for the culture of those germs; might it not, indeed, be the 'intermediate host' necessary for some phase of their development?"[6]

Examination of the context, however, shows that Finlay at the time was thinking of a process by which the amount of morbid material might be increased. In his view, an attack ought to occur after several bites or after one bite by a mosquito infected some time earlier, for in the latter case "the germs have had time to develop more abundantly . . . and the virulence of the inoculation might be expected to become proportionately increased."[7] On the basis of this argument he answered Corre's objection that one bite was not enough to produce a serious case of the disease. The same reasoning also accounted for the Saint Nazaire epidemic of 1891: the germs had had time to multiply during the crossing. But the assumed equivalence between one bite by a mosquito infected some time earlier and several bites by a recently infected mosquito was incompatible with the idea of a cycle. The assumption was compatible, however, with the normative idea that a single bite by a recently infected mosquito defined the minimal dose needed to induce a mild fever and immunity to the disease.

Finlay was fully aware that his theory was based on the view that the mosquito resembled a graduated syringe. This knowledge demolished his claim that he had anticipated the concept of host in the sense that term acquired as a result of Ross's work. No doubt that was his reason for issuing a correction, by inserting his 1886 statement into a new context in which the notion of cycle appeared. In 1902 Finlay claimed that this idea had occurred to him on reading "an account published in Van Tieghem's *Botanique* (1884 edition, p. 1035) of the life-cycle of the *Puccinia graminis*, in which I was much interested."[8]

We should not be misled by this information about Finlay's supposed inspiration. If the analogy with the cycle of *Puccinia graminis* had really been that illuminating, Finlay would have cited Van Tieghem in 1886. In reality, Finlay took the term *host* from Manson and used it as early as 1882 in speaking of filaria: "As far as this microzoon is concerned, it does not seem possible that inoculation takes place directly; rather, it must proceed by way of an intermediate host, as some English physicans have argued."[9]

The reference to the *Traité de botanique* offered two advantages. First, it allowed Finlay not to say that he had borrowed the expression from Manson. Second, it established, in hindsight, the consistency of his thinking, since Finlay based the alleged heuristic function of the analogy on his hypothesis of the moment (that the etiological agent was a fungus). Furthermore, Finlay himself later disposed of this misleading version by stating several times that he had not considered the idea of a host until much later: "Since [*sic*] 1898, I formulated the principle that . . . the germs would multiply in the body of the insect and finally invade its salivary and venom glands."[10]

Reed, for his part, attempted to discredit Finlay's hypothesis completely. He tried to demonstrate his priority by showing that the idea of exploring the mosquito theory was his own. Reed knew full well, however, that Finlay had given the clue. In order to get around this obstacle, Reed insisted on the Cuban physician's failures: "If, indeed, one were guided by the result obtained, the only logical conclusion to be drawn was that Finlay had disproved his own theory."[11] Since truth excludes falsehood, it was apparently self-evident that the Americans had not taken Finlay's theory into account.

Hence another point of departure had to be proposed. It was therefore alleged that when the American commission had arrived in Havana, it first considered the *fomites* theory, which, Reed claimed, was disposed of by means of observations made at Pinar del Rio. This, Reed argued, led to the idea that the commission ought to consider "some insect capable of conveying the infection, such as the mosquito. . . . This was, however, only a supposition."[12] Examination of the context, however, shows that this version is scarcely credible. At the time the commission's attention was focused on Sanarelli's bacillus.

Reed was well aware that before arriving in Havana the commission had not paid the slightest attention to the problem of transmission. After the fact, therefore, considerable attention was paid to what exactly had aroused Reed's interest in the mosquito theory: "You [Carter] must not forget that your work in Mississippi did more to impress me with the importance of an intermediate host than everything else put together."[13] The following month Reed added Ross's work to the list: "If malarial fever . . . required the agency of a special germ of mosquito for its propagation . . . it did not seem unreasonable to suppose that yellow fever . . . might also depend on some agent for its spread."[14]

Make no mistake, however: if the work of Carter and Ross had been so illuminating, Reed would have started research in that direction in June. In fact, it was the English who provided the missing link in July, and their hint led to Finlay's hypothesis. There were two advantages to mentioning only the work of Carter and Ross. First, Reed was able to omit the names of Durham and Myers, who had initiated the correct chain of reasoning. Second, he was able to omit the last link in that chain: Finlay. The American commission's visit to Finlay suffices to refute Reed's claims.

In short, Finlay and Reed argued in symmetric and opposite ways. Their behavior amounts to proof by contradiction of an axiom of historical epistemology: both men sought to establish the validity of a single point of view and to analyze the past in terms of truth and error. Finlay attempted to set his hypothesis in the context of scientific truth and thus to show that he had anticipated the Americans. Meanwhile, Reed pointed out that Finlay's hypothesis was false and attempted to show that the commission's work did not overlap that of his Cuban rival. Both Finlay and Reed did what they did in order to ensure for themselves a place in history. Finlay feared that his work would be unjustly dismissed as superseded, while Reed feared that his role would be unjustly reduced to that of verifying someone else's theory. These fears were legitimate. The way in which Finlay and Reed tried to dispel them was less so.

These first interpretations established a way of looking at the history of this discovery that has persisted for a long time. By demonstrating that he had anticipated the Americans' conclusions, Finlay suggested a close connection between his work and the commission's. Conversely, by showing that the commission's research did not depend on Finlay's, Reed suggested a sharp break. Finlay bequeathed to historians a principle of continuity, Reed a principle of discontinuity.

LEGENDARY HISTORIES

Seduced by these interpretations, historians have based their accounts on them. In so doing they have tended to sharpen the differences between the two versions by choosing to emphasize either continuity or discontinuity, or, to put it another way, by focusing on either the precursors or the founders of the final result. Cuban historians have shown that Finlay did anticipate Reed. Since Reed said nothing Finlay had not said, Finlay was the founder. By contrast, American historians have shown that Finlay had a whole constellation of predecessors. Since Finlay was no innovator, Reed was the founder of the result. All discourse can thus be characterized as either original or repetitive. Historians have assigned themselves the task of distinguishing the former from the latter. At the same time they have inevitably celebrated the genius of the founders whose work inaugurated the period of genuine science.

Cuban historians claim that the Americans simply copied Finlay's work. To do so they simply invoke Finlay's own argument that he was the source of everything: "One after another, the conclusions reached by the American commission simply reproduced Finlay's concepts and ideas in every detail."[15] What led the commission to undertake its work of verification? Critical analysis of the American effort clarifies this point by recalling, first of all, the context: indifferent to Finlay's theory, the commission busied itself with Sanarelli's bacillus. If eventually it came to the same conclusions as Finlay, it was in spite of itself, indeed in refutation of its own view. In effect, the commission carried out its experiments "not so much to assist in the official verification of a new discovery for the sake of science as to confirm Dr. Finlay's failure."[16] Everyone knows what happened next: the unbelievers came down with the very disease they were supposed to be studying. The verification was *accidental*.

To show that Finlay anticipated Reed is to reduce the commission's role to one of mere execution: it approved Finlay's ideas and communicated them to the scientific community. To focus on the way in which the commission conducted its work is to say that Finlay was to the Americans as the true believer in a scientific theory is to those who jump on the bandwagon at the last minute. The Cuban historians extol Finlay by discrediting the American commission.

The result has been legend in the guise of history. Finlay was a man of exceptional gifts. "From the Scotch he inherited the intensity and the ardor in pursuit of an objective. . . . From the French he received . . . a lively imagination."[17] These talents needed an occasion in order to reveal themselves, however. Two accidents serve as the linchpin in this account of the theory's history: the visit of the Chaillé Commission and the reading of Van Tieghem's *Traité de botanique*. The commission provided Finlay with the problem and Van Tieghem with the solution: "In a happy coincidence, while [Finlay] was reflecting on these matters and looking for the agent mentioned by the American commission, Van Tieghem's book on botany came into his possession, and on page 1035 he read the description of the life cycle of *Puccinia graminis*. . . . It occurred to Finlay that what took place in plants might also take place in yellow fever."[18] Everything else follows: the analogy led to the hypothesis, which was immediately verified by experimental inoculations. Finlay's work marks a threshold of scientificity.

For American historians the first order of business was to locate Finlay's predecessors. In the respect Reed, Carroll, and Agramonte behaved, if I may put it this way, as precursors. As long ago as the turn of the century they exhumed Nott, Dowell, Crawford, and Beauperthuy—so many names to choose among. I shall quote from just one brilliant synthesis of this work: "Slightly anticipating Carlos J. Finlay . . . were reports from Texas suggesting the possible influence of mosquitoes in the transmission of the disease. Greensville Dowell (1822–1881) deserves recognition with John Crawford (1746–1813), Josiah C. Nott (1804–1873), and Louis D. Beauperthuy (1807–1871) as one of those pioneer observers who associated mosquitoes with yellow fever epidemics."[19] Critical analysis of Finlay's work is limited to pointing out that he changed his theory constantly and above all that he failed to prove his case: "The experiments possess absolutely no weight; they are entirely unscientific."[20]

To list Finlay's precursors is to show that his hypothesis was essentially nothing more than an old *belief* with no real scientific importance. Critical analysis of Finlay's work was aimed at revealing a project symmetrical with but opposite to that of Reed: Finlay is to Reed as error is to truth, as uncertainly is to certainty, as ineptness is to skill. In depreciating Finlay's work American historians were of course praising Reed's. When the commission arrived in Havana, it found itself confronting a "mystery."

Again the result was legend in the guise of history. The problem could only have been solved, it was said, by "the possessor of a striking personality, unusual reasoning power, and unbiased judgment."[21] Fervent biographers spoke of "Reed's genius." But genius can reveal itself only where truth passes unnoticed by ordinary mortals. An object must satisfy two conditions to be worthy of the genius's notice: it must be sufficiently remote to allow room for the display of talent yet sufficiently pertinent to shed new light on matters of interest. Carter's work satisfied both conditions: "A year before another Virginian, Dr. Henry R. Carter . . . made an elaborate statistical study of the spread of yellow fever in houses. . . . [This was] one of the most important scientific works which had been done up to that time on yellow fever, and in itself furnished the starting point for Reed's investigations."[22] Everything else follows: from Carter's observations came the correct hypothesis, which was promptly confirmed by experimental inoculations. Reed's work marks a threshold of scientificity.

Historians will be slow to give up their legends despite their errors. It may be worth mentioning a few pertinent facts. In order to see Finlay as Reed's predecessor, one must forget the fact that Ross's work constitutes a watershed. In order to see Beauperthuy as Finlay's predecessor, one must forget the difference made by Manson's work. What accounts for these misreadings? The mistaking of a word for a concept, without which no precursors would exist. Just as Cuban historians seize on the expression "intermediate host" in Finlay's work as if it contained the concept, so, too, do American historians seize on the word "mosquito" in Beauperthuy as if it contained the notion of an agent of transmission. Finlay, as we have seen, was referring to a process that might increase the quantity of yellow fever toxin and not to any kind of cycle. And Beauperthuy was talking not about the transmission of germs from man to man but about the transport of a poison from swamp to man: "The accidents of yellow fever, in my view, also have to do with the introduction into the [organism's] economy of septic juices imbibed along the coast by insects."[23]

When Cuban historians charge that the members of the American commission were incredulous opportunists who carried out their experiments with mosquitoes for the sole purpose of discrediting Finlay, they neglect the crucial point: namely, that the Americans understood the need for the germs to incubate for a period of time in the insect's body. Similarly, when American historians see Finlay's work as a baseless belief whose content was constantly shifting, they, too, obscure an important point: namely, that the Cuban physician was the first to point a finger of guilt at the mosquito as the agent of transmission. Both sets of historians have only a hazy view of the object under study: psychology has absolutely nothing to do with the rules governing the formation of hypotheses and their experimental verification.

Finally, to single out the heuristic function of the analogy with *Puccinia graminis* is to substitute Van Tieghem for Manson. Similarly, to insist on the importance of observations pertaining to the period of extrinsic incubation is to substitute Carter for Ross. In both cases the result is a misinterpretation. The Cuban historians fail to notice that the chronology is inflexible. How can the *Traité de botanique*, published in 1884, be invoked to explain a discovery made in 1881? Perhaps there was some other source of information that Finlay forgot to mention. That is not possible. In 1882 a controversy arose at the Academy of Sciences in Havana concerning a disease in coconut palms caused by a fungus belonging to the genus *Puccinia* of Tulasne. If Finlay knew that wheat rust could be halted by eradication of barberry plants, he no doubt would have asked himself if the coconut parasite undergoes a transformation analogous to that of *Puccinia graminis*. Instead he stated that he was "of the opinion that the diseased coconut palms must be cut down . . . because neither the contagion factor nor its mode of transmission is yet known with certainty."[24] The American historians, for their part, have failed to notice that Carter's observations take on meaning only in light of Ross's work on malaria.

RELATIONS OF POWER

In 1900 the commanding general of the occupation army organized a banquet in Finlay's honor. This occasion was the first public homage to the Cuban physician and the first celebration of a famous collaboration. Politically it was a shrewd move. It flattered the pride of the Cuban elite, satisfied the members of the commission of inquiry, and retrospectively legitimized the American intervention in Cuba: "The confirmation of Dr. Finlay's doctrine is the greatest step forward made by the medical sciences since Jenner's discovery of the vaccination; this accomplishment alone justifies the war against Spain."[25]

Not until 1955 were the spirit and warmth of the 1900 banquet to be seen again. Not for lack of opportunities, however. In 1933, for example, on December 3 (declared "American Medical Day" by the Pan-American Association), the United States might have joined with Cuba in celebrating the centennial of Finlay's birth. By 1955, however, circumstances were auspicious for a new—and novel—celebration, also a centennial of sorts: the Jefferson Medical College of Philadelphia had awarded Finlay his medical degree in 1855.

The homage was not without ironies to savor. The Cuban dictator, Batista, came in the guise of a patron of public health and was rewarded with an honorary doctorate in humanities. The Americans, meanwhile, were able to stress the role of the United States in training the Cuban physician. A disciple of Silas Weir Mitchell, it was no doubt in Philadelphia that Finlay had acquired the intellectual qualities that had allowed him to make his discovery. A hundred years after his studies at Jefferson Medical College, Uncle Sam afforded him the privilege of becoming "a famous adopted son."[26] This commemoration gave brief if cynical expression to the imperialist dream, to that "good neighbor policy" that was an impossibility even in science—the exception that confirms the rule.

In fact, when the Americans did not simply ignore Finlay's work they vigorously raised the controversial issue: "He [Finlay] is today acclaimed throughout all Latin America as the real hero of the Mosquito–Yellow Fever discoveries. With this, however, North America . . . does not agree. Since he did not prove his case."[27] By "discovery" the Americans meant the designation of the mosquito as intermediate host, the key to producing an experimetal case. By "discovery" the Cubans meant the selection of the right species. When it was proposed that the mosquito be named *Stegomyia finlayensis*, Theobald, an authority on classification, observed that if the law of priority were to be respected it would have to be named *Stegomyia fasciatus fabricus*. The dispute was irresolvable. Each side emphasized what it took to be the salient feature of the discovery: the theory was either reduced to the capture of the mosquito or replaced by the experimental manipulation of the insect. There was ample reason for this blindness: only if the theory was indentified with either Finlay or Reed could either man be taken as a symbol, respectively, for Cuba or the United States.

The controversy was hardly limited to the question of priority. It is wrong to assume that it continued to rage through much of the twentieth century because the two positions were irreconcilable. The controversy, ostensibly a dispute over scientific credit, actually reflected a struggle for power. The Americans pressed Reed's claim in order to conceal Finlay's role and establish the supremacy of "triumphant" American medicine over that of a dominated nation. The Cubans pressed Finlay's claim in opposition to Reed in order to demonstrate their resistance to American ascendancy over Cuba. This struggle was quite unlike other by-products of ethnic and cultural domination. Reed's name, to put it mildly, offended Cuban ears: it was "one of the most flagrant expressions of the arrogant provocation and injustice that many North Americans owing to an illegitimate sense of superiority manifest toward Cuba."[28]

In order to understand this controversy we must briefly recall what led up to it. Before Finlay and Reed became symbols, despotism and yellow fever were seen as twin scourges, a common enemy: "We are fighting not only for the freedom of Cuba from Spanish tyranny, but for the freedom of America from Spanish disease."[29] This conviction, which inspired American democrats, may have been shared by Cuban patriots. In any case, it is certain that the United States and Cuba were joined together in a battle against old-world oppression. At the turn of the century the two countries jointly celebrated their double victory: not only had Spain been defeated, but yellow fever would soon be eradicated as well. Success in the field of public health followed hard on the heels of a military, not to say a humanitarian, victory. For a brief instant Cubans believed that the McKinley administration had hastened the day when the dreams of Martí, martyred five years earlier in the battle of Dos Rios, would be realized. For a brief instant they also believed that Reed had made Finlay's dream a reality: the eradication of yellow fever through destruction of the mosquito.

It soon became clear, however, that McKinley's successor, Theodore Roosevelt, did not share Martí's, ideals. Imperialism was already the great issue of the day. The Platt amendment, which placed Cuba under the safeguard of the United States, humiliated the Cuban people and opened their eyes to the meaning of the American intervention: the United States had supplanted Spain. Then and only then did the names Reed and Finlay become convenient, incontrovertible symbols of a clash of values. Reed was no longer just the "conqueror" of yellow fever; now he also embodied the ambitions of Roosevelt. Reed was one of the most prestigious promoters of the great work of pacification, the man responsible for "the advance of commerce and civilization, the rejuvenation of Latin America."[30] And Finlay was no longer just the author of a revolutionary theory but also the embodiment of Martí's ideal, one of the most illustrious exemplars of the Cuban genius. His name epitomized an oppressed people's claims to dignity.

Thus the mythical version, which was also a profoundly political version, of the discovery was put to use by society. The myth was historical knowledge in the form of instruction, or, in other words, a social use of the historical message. That message was spread in various ways. It occupied historic sites, flowed through channels of information, and was trumpeted to the masses by theater, film, and television.

Consider first the occupation of historic sites. In 1905 a marble bust of Walter Reed was placed in Washington's National Museum. In 1916 a monument in honor of Finlay was unveiled in the garden of Cuba's Ministry of Public Health. These expressions of gratitude did not come close to satisfying the vanity of nationalists, however. In the United States Hemmeter proposed a new monument "of the Genius of History presenting to American Medical Science the laurels of immortality, commemorating the discoveries and martyrdom of the United States Yellow Fever Commission."[31] In Cuba it was proposed that a majestic and severe statue of Finlay be erected at the entrance of the harbor "so that every traveler and immigrant may know that, if Columbus discovered America, Finlay made it habitable."[32]

Soon, however, these grandiose projects had to be abandoned in favor of more prosaic tasks. Here I shall mention only the most notorious incident. New York University's Hall of Fame contained busts of Reed and Gorgas but none of Finlay, a fact that was taken as denying "the glory due the great Cuban and the West Indian nation in which he made his extraordinary discovery."[33] In 1956 W. Duvon Corbitt, "friend of Cuba," launched a campaign to have Finlay included. Rotary Clubs were mobilized in this effort, ostensibly because part of their mission was to bring peoples closer together. In 1960 hopes were dashed when Finlay was denied admission to the Hall of Fame on the grounds that he was not an American citizen.

The message was also transmitted through the usual informational channels. In *American Medical Biographies* one finds this entry: "Reed, Walter (1851–1902). Walter Reed, chairman of the United States Army Yellow Fever Commission and discoverer of the mode of propagation of the disease."[34] Rodriguez Exposito would have put it quite differently. Under the Cuban's name he would have written: "Finlay: Cuban physician, born at Camagüey (Cuba) on December 3, 1833, and responsible for the discovery of the mode of propagation of yellow fever."[35] Beyond that, he would have erased the error wherever it appeared: in textbooks, popularizations, dictionaries, and encyclopedias—an impossible task. Still, it was essential to remain vigilant at all times. And Cuban historians have tried. Consider one example. In 1910 Osler gave a lecture at the School of Tropical Medicine in London in which Finlay's name was not mentioned. Guiteras was quick to respond: "He [Finlay] also participates, on equal terms, I think, in the great glory of the discovery, thereby making it that much less of an exclusive American triumph."[36]

Finally we come to the mass media. In 1934 *Yellow Jack*, a play written by Sidney Howard and Paul De Kruif, author of the indescribable best-seller *Microbe Hunters* (1928), caused a commotion in New York. Metro-Goldwyn-Mayer soon brought the play to the screen. Cuban filmgoers were outraged: ultimately Finlay does appear in the film, but capriciously distorted, "a Finlay decked out with a monocle and with the sideburns of an old sea captian, which he never wore in his life. As for his scientific discovery, no clear recognition of his achievement was permitted since our compatriot was forced to play second fiddle to Dr. Reed."[37] A response was quick to appear: Roldan wrote a screenplay in five acts about the life and work of the illustrious Cuban. The scenario traced the history from Finlay's discovery to the final eradication of the disease that it made possible.

To conclude, mention should be made of one notorious—and deplorable—incident, which unleashed a tremendous amount of passion. On the television quiz show "The $64,000 Question," a contestant was asked who conquered yellow fever. The winning answer was Walter Reed. This time Cuban indignation was so intense that El Caudillo Batista himself stepped into the arena to claim that the discovery was of course due to Finlay. Steven Carlin, the program's producer, tried to make amends. He was especially appreciative of Cuban interest in his program, but he was also willing to acknowledge Finlay's claim to fame.

Only the award of a prize could have resolved the dispute. In 1907 Finlay received the Mary Kingsley Medal, but Gorgas had already won it. There was talk of a Nobel Prize. In 1904 Ross proposed Finlay. The American commission also had its supporters. In 1906 Howard and Ross discussed the possibility of nominating Carroll. Thought was also given to splitting the prize: in the United States

Carroll and Agramonte were suggested; in Cuba, Agramonte and Finlay. These were certainly odd couples: the first would have insulted the Cubans, while the second would have ruffled the sensibilities of the Americans. Finlay's backers then besieged the International Congress of the History of Medicine. It proved to be a receptive audience. In 1935 the Cuban delegation persuaded the Tenth Congress at Madrid to recognize Finlay's primordial role and to denounce the American usurpation. In 1954 the Cubans persuaded the Fourteenth Congress at Rome to ratify the 1935 accord a second time. In 1956 it was ratified a third time by the Fifteenth Congress at Madrid. (The "$64,000 Question" incident, it should be noted, took place in 1955.) "Once again we confirm that the discovery of the agent of transmission of yellow fever was made by Carlos Finlay and by him alone."[38]

CONCLUSION

In examining scientific discourse, we have had to distinguish between what scientists said in reflecting on their own practice and what is actually relevant to the history of scientific ideas. When scientists reflect on their practice, they offer a distorted narrative: a contingent history that highlights their perspicacity and decisiveness. In describing these accounts, my purpose was not to redistribute merits or to diminish the originality of one man so as to increase that of another. It was rather to indicate the function of these divergent versions of history and to eliminate the confusion they have produced. I wanted to show how and why Finlay concealed Manson's influence and the American commission concealed the role of the English in order to clear the way for another view of history. In other words, I have attempted to lay the groundwork for the history of science in the true sense: namely, the analysis of theoretical structures and scientific propositions, of conceptual building blocks and their field of application.

Showing how Finlay borrowed and made use of Manson's theory made the formulation of his hypothesis sensible and comprehensible. Similar, showing how Durham and Myers pointed the American commission in the right direction revealed a whole range of necessities behind this one minor event. My objective was to describe those medical methods that were actually employed and the rules that governed their use. From a work such as this Finlay and Reed emerge neither magnified nor diminished. The merit of individuals is not at issue.

In our examination of scientific discourse we have also had to distinguish between propositions that were simply false and propositions that were ultimately superseded as a result of epistemological transformations. When scientists (or historians) engage in critical epistemology, they point out the errors of their predecessors and mark a dividing line that inevitably separate what is not yet scientific from what is truly science. In describing this level of discourse, my aim has not been to repeat it: to discover what was fruitful in error and thus show how erroneous yet subtle intuition might indeed be worthy of admiration. My purpose, rather, was to show how scientists' analyses of their own work function in order to avoid their inherent reductionism. I wanted to show how and why Ross concealed the historical significance of Manson's work, and Reed that of Finlay's. In other words, I hoped to clear the way for another view of history. Or, to put it another way, I hope to lay the groundwork for a true archaeology of science: that is, an analysis of the transformation of fields of knowledge, an identification of epistemological thresholds, an examination of how objects, concepts, and theories are formed.

To describe the transformations that took place in

tropical medicine was to show how false propositions may give rise to a discursive practice out of which new propositions can take shape. Those transformations revealed a new object of study, the arthropod; new disciplines, such as comparative parasitology, experimental parasitology, and medical entomology; and new ideas, such as that of an active intermediate host.

Because epidemiology called into play a range of novel procedures including recording of cases, tracing of interdependencies, and study of the biology of organisms, it was transformed into a "natural history of transmissible diseases."[39] By contrast, classical epidemiology was something more in the nature of a cultural history of transmissible diseases,[40] and its liberation allowed the existence of "pathogenic complexes" to emerge.

The analysis in this book rests on a narrow historical base: it deals with the development of epidemiology and associated methods over a period of barely twenty years. That period, however, happens to have been one that established a permanent watershed. It was a time when the hosts of such little-known disease as filariasis and such notorious diseases as malaria and yellow fever were revealed. A nematode, a protozoon, and a virus each came to occupy its respective host. Something that was fundamentally invisible was finally understood throughout the complexity of its interactions. One has the impression that for the first time physicians swerved from the beaten path to follow the strange ways of destiny. But the order of the analysis should be reversed: it was the forms of visibility that changed. The new epidemiological spirit first evinced by the work of Manson, Laveran, and Finlay was not the result of some sort of psychological or scientific purgation. Before this new form of knowledge could come into being, observation had to be restructured, new objects had to be perceived, and the relations among living things had to be redefined.

These three diseases (elephantiasis, malaria, and yellow fever), which might well be termed baroque because of their symptoms, proved to be even more baroque in their modes of transmission. The late nineteenth century thus discovered strangely machinelike interworkings among living things. Disease broke out where living things interacted while pertinaciously pursuing their own ends. Whereas contemporaries were content simply to retrace the processes of pathology, Manson, Laveran, and Finlay were the first to map the pathological world. And of course mapmaking is the opposite of tracing: "[A map] is an experiment performed directly on reality. . . . It contributes to the connection of fields, to the removal of obstructions, to a maximization of openness against a background of uniformity."[41]

After Laveran, Manson, and Finlay, a very old symbol was drained of its meaning: death came now not in the form of a man with a scythe but of a biting insect. The needle supplanted the scythe.

Notes

2 Multiplicities

1
R. Dean, *Report of Yellow Fever in the U.S.S. Plymouth in 1878–9* (Washington, 1880), pp. 66–67.

2
Ibid., p. 35.

3
Ibid., p. 33.

4
Ibid., p. 43.

5
Ibid., pp. 35–36.

6
Ibid., p. 39.

7
Ibid., p. 34, note 1.

8
S. Chaillé, *Conclusions of the Board of Experts Authorized by Congress to Investigate the Yellow Fever Epidemics of 1878* (Washington, 1879), p. 9.

9
Dean, *Report on Yellow Fever*, p. 47.

10
S. Bemiss, "Chapters from Report of Yellow Fever Commission of 1878. Nature of Yellow Fever," *New Orleans Medical and Surgical Journal* 10(1882):326.

11
Cited in P. Manson-Bahr and A. Alcock, *The Life and Work of Sir Patrick Manson* (London, 1927), pp. 51–52.

12
T. Cobbold, "The Life-history of *Filaria bancrofti*, as explained by the discoveries of Wucherer, Lewis, Bancroft, Manson, Sonsino, myself, and others," *Journal of the Linnean Society: Zoology* 14(1879):366–367.

13
P. Manson, "Further Observations on *Filaria Sanguinis Hominis*," *Medical Reports. China Imperial Maritime Customs* 2(1877):9. The key portion of this paper was published in England the following year under a different title: "On the Development of *Filaria sanguinis hominis*, and on the Mosquito Considered as a Nurse," *Transactions of the Linnean Society: Zoology*, 1878, pp. 304–311.

14
P. Manson, "Lymph Scrotum, Showing Filaria *in situ*," *Transactions of the Pathological Society* 32(1881):290. My preference is to cite this paper wherever possible; it is the only one in which Manson explained his method of research.

15
Manson, "Further Observations," p. 11.

16
Ibid., p. 13. In 1877 Manson traced the embryo's development in the abdomen. In 1884, following and based on observations of Lewis, Manson added a description of the migratory phase toward the thorax. This minor modification does not justify the claim that Manson completed his theory in 1884. Compare "The Metamorphosis of *Filaria sanguinis hominis* in the Mosquito," *Transactions of the Linnean Society: Zoology* 2(1884):367–388.

17
Ibid., p. 12. On this point Manson's idea would not change. Compare Manson, "Lymph Scrotum," pp. 292–293.

18
Ibid., p. 14. Manson was here borrowing prevailing ideas concerning the filaria's mode of entry. Compare C. Davaine, *Traité des entozoaires* (Paris, 1877), p. 795: "In most areas where filaria is prevalent, it is widely believed that it is taken in along with water."

19
S. Chaillé and G. Sternberg, "Preliminary Report of the Havana Yellow-Fever Commission," *Annual Report of the National Board of Health* (Washington, 1879), p. 34. The final report was published the following year: "Report of Havana Yellow-Fever Commission," *Annual Report of the National Board of Health* (Washington, 1880), pp. 64–308.

20
Ibid., pp. 64–65.

21
Ibid., pp. 70–71.

22
Ibid., p. 66.

23
Ibid., p. 69.

24
Ibid., p. 71. See Carlos Finlay, "La etiología de la fiebre amarilla," 1858, *Obras Completas,* 6 vols. (Havana, 1965–1983; hereafter cited as *OC*), vol. 1, pp. 87–118; "Alcalinidad atmosférica observada en La Habana," 1872, ibid., pp. 119–126; "Carta sobre alcalinidad atmosférica," 1873–1874, ibid., pp. 139–141.

25
Ibid., p. 56.

3 The Formulation of a Hypothesis

1
C. Finlay, "Conferencia Sanitaria Internacional de Washington," 1881, *OC,* vol. 1, p. 195 (referred to in the text as the 1881 memoir).

2
Da Silva Amado, "La conférence sanitaire internationale de Washington," *Revue d'hygiène et de police sanitaire* 3(1881): 378.

3
C. Finlay, "Fiebre amarilla. Estudio clinico patológico y etiológico," 1895, *OC,* vol. 2, p. 162.

4
F. Dominguez Roldan, *Docteur Carlos J. Finlay, son centenaire (1933), sa découverte (1881)* (Paris, 1935), pp. 45–46. This version of history is repeated in H. Scott, *A History of Tropical Medicine* (Baltimore, 1939), vol. 3, p. 1031, and C. Singer and E. Ashworth Underwood, *A Short History of Medicine* (New York and Oxford, 1939), p. 469.

5
F. Hurtado Galtes, H. Abascal y Vera, C. Rodriguez Exposito, "Finlay en la historia de la medicina," *Cuadernos de historia de la salud pública* 7(1954):35.

6
G. Delgado Garcia, "La doctrina finlaísta: valoración científica e histórica a un siglo de su presentación," *Cuadernos de historia de la salud pública* 67(1982):15.

7
Cobbold, "The Life-history of *Filaria bancrofti*," p. 346.

8
"Discusión del 'informe sobre secuestración de los lazarinos' del doctor Emiliano Nuñez," 1879, OC, vol. 3, p. 560.

9
"Discusión del 'informe sobre secuestración de los lazarinos' del doctor José I. Torralbas," 1880, OC, vol. 3, p. 574.

10
J. Fayrer, "On the Relation of *Filaria Sanguinis Hominis* to the Endemic Disease of India," *The Lancet*, 1879, p. 221a. In Cobbold, "Discovery of the Intermediary Host of F.H.S., (F. Bancrofti)," *The Lancet*, 1878, p. 69b, we read: "If medical men will only reflect to what extent diseases hitherto obscure (such as chyluria, elephantiasis, lymph-scrotum, and perhaps leprosy itself) are henceforth to be associated with the bite of a gnat . . ."

11
"Pathological Society of London," *The Lancet*, 1879, p. 268a.

12
C. Finlay, "Consideraciones acerca de algunos casos de filariasis observados en La Habana," 1882, OC, vol. 4, p. 22.

13
Ibid., p. 24. Incidentally, a collection of *The Lancet* can still be found in the library of Havana's medical school.

14
C. Finlay, "¿Es el mosquito el único agente de transmisión de la Fiebre Amarilla?," 1902, OC, vol. 3, p. 97.

15
Finlay, "Consideraciones," p. 22.

16
Finlay, "El mosquito hipotéticamente considerado como agente de transmisión de la fiebre amarilla," 1881, OC, vol. 1, p. 248. Finlay's silence was not accidental. Consider another example: "In December 1880 these considerations (and others which I omit for reasons of brevity) led me to think that the unique mode of transmission . . . must be inoculation." See "Fiebre amarilla. Estudio clinico patológico y etiológico," 1894, OC, vol. 2, p. 162.

17
Finlay "Reseña de los progresos realizados en el siglo XIX en el estudio de la propagación de la fiebre amarilla," 1901, OC, vol. 2, p. 12.

18
Finlay, "El mosquito," OC, vol. 1, p. 257.

19
"The Mosquito Hypothetically Considered as the Agent of Transmission of Yellow Fever," *New Orleans Medical and Surgical Journal* 9(1882):611.

20
C. Finlay, "Excerpt from the Report," in "Report of the Havana Yellow-Fever Commission," *Annual Report of the National Board of Health* (Washington, 1880), p. 165.

21
C. Finlay, "Transmisión del cólera por medio de las aguas corrientes cargadas de principios específicos," 1873, OC, vol. 3, p. 398.

22
S. Ffirth, *A Treatise on Malignant Fever with an Attempt to Prove Its Non-Contagious Nature* (Philadelphia, 1804), pp. 46–60. Finlay probably learned of Ffirth's experiments from the classic work of R. La Roche, *Yellow Fever, Considered in Its Historical, Pathological, Etiological, and Therapeutical Relations*, 2 vols. (Philadelphia, 1853). For Ffirth's conclusions see the detailed article of S. Peller, "Walter Reed, C. Finlay, and Their Predecessors around 1800," *Bulletin of the History of Medicine* 32(1959):196–211.

23
Finlay actually wrote "zoofitos." The correction was inevitable, because the hypothesis concerned an agent of transmission. It was therefore absurd to specify an immobile organism. The next sentence and the context, moreover, confirm that he was indeed talking about microphytes: "These lowest orders of animal life . . . "

24
Finlay, "El mosquito," p. 248.

25
C. Finlay, "La fiebre amarilla, su transmisión por intermedio del *Culex mosquito*," 1886, OC, vol. 1, p. 420.

26
J. Jones, "Yellow Fever Epidemic of 1878 in New Orleans," *New Orleans Medical and Surgical Journal* 6(1879):713.

27
R. Blanchard, *Les moustiques, histoire naturelle et médicale* (Paris, 1905), p. 507.

28
C. Finlay, "Letter to Sr. D. Le Roy de Mericourt," 1884, OC, vol. 1, p. 294. Finlay resurrected this idea on several occasions: "The characteristic lesion of yellow fever is located in the walls of the

vessels, whose tissues are inevitably traversed by the proboscis of the mosquito at the time of the bite." See C. Finlay, "Fiebre amarilla experimental comparada con la natural en sus formas benigna," 1884, *OC*, vol. 1, p. 321. In 1888 Finlay claimed to have observed the microbes in the capillary walls; in 1891 he repeated this claim: "It is not from the blood that has been sucked that the mosquito is supposed to derive its contamination but from the tissues that the sting must bore through." See "Transmission of the Yellow Fever by the *Culex mosquito*," 1891, *OC*, vol. 3, p. 184.

29
Finlay, "El mosquito," pp. 257–258.

30
P. Manson, "A Short Autobiography," *Journal of Tropical Medicine and Hygiene*, 1922, p. 159b.

31
P. Manson, "The Metamorphosis of *Filaria sanguinis hominis* in the Mosquito," *Transactions of the Linnean Society: Zoology* 2(1884):383.

32
Finlay, "El mosquito," p. 247.

33
P. Manson-Bahr, "The Story of Malaria: The Drama and Actors," *International Review of Tropical Medicine* 2(1963):331.

34
Finlay, "El mosquito," pp. 252–254.

4 Bacteriological Research

1
D. Freire, *Doctrine microbienne de la fièvre jaune et ses inoculations préventives* (Rio de Janeiro, 1885), p. 247.

2
C. Finlay and C. Delgado, "Estado actual de nuestros conocimientos tocante a la fiebre amarilla," 1887, *OC*, vol. 1, p. 447.

3
Freire, *Doctrine microbienne*, p. 267.

4
Ibid., p. 17. Freire also cites the article "Algue" in Littré and Robin, *Dictionnaire de Médecine, Chirurgie et Pharmacie*, 1873, pp. 44–45.

5
J. B. Lacerda, "The Parasitic Origin of Yellow Fever," *The Sanitarian* 11(1883):612b.

6
J. B. Lacerda, "De la cause primordiale de la fièvre jaune," *Gazette des Hôpitaux* 56(1883):821b. See also "Mémoire présenté par M. de Quatrefages," *Comptes rendus hebdomadaires de l'Académie des Sciences* 96(1883):1708–1709.

7
M. Carmona y Valle, *Leçons sur l'étiologie et la prophylaxie de la fièvre jaune* (Mexico, 1885), p. 202. The first paper dates from 1881: "Estudio etiológico de la fiebre amarilla," *Gaceta Médica de México* 16(1881):385–401.

8
Ibid., p. 92.

9
C. Finlay, "Nuevos datos acerca de la relación entre la fiebre amarilla y el mosquito," 1883, OC, vol. 1, p. 302.

10
C. Finlay and C. Delgado, "Estudio actual de nuestros conocimientos tocante a la fiebre amarilla," 1887, OC, vol. 1, p. 448. See also "Del micrococo tetrágeno de la fiebre amarilla," 1888, ibid., vol. 2, pp. 11–18; "Relación entre los factores etiológicos y la evolución de la fiebre amarilla," 1888, ibid., vol. 2, pp. 31–36; "Resumen de nuestras investigaciones sobre tetrágenos en la fiebre amarilla," 1888, ibid., vol. 2, pp. 37–44; "Communicacion sobre el *micrococcus Febris Flavae*," 1888, ibid., vol. 2, pp. 49–51; "Resumen de nuestras investigaciones sobre etiología de la fiebre amarilla en el año de 1888 a 1889," 1889, ibid., vol. 2, pp. 53–63; "Resultado de los experimentos comparativos hechos sobre el *Micrococo Tetragenus versatilis*," 1889, ibid., pp. 65–73.

11
"Rapport de M. Rochard sur une note de Freire intitulée Etudes expérimentales sur la contagion de la fièvre jaune," *Bulletin de l'Académie de Médecine* 13(1884):576.

12
Ibid., p. 583.

13
"Carmona's Yellow Fever Inoculations," *New Orleans Medical and Surgical Journal* 13(1886):972. Concerning the debate, see "The Yellow Fever Commission," ibid., pp. 547–571, 722–728, and 964–994. See also "The Proposed Yellow Fever Commission and the National Board of Health," *Atlanta Medical and Surgical Journal* 3(1887):178–183; "The Yellow Fever Commission," *Journal of the Medical Association* 6(1886):307b. J. Holt set forth his views in "Yellow Fever Commission," *New Orleans Medical and Surgical Journal* 13(1886):625. On the favorable

reception of Freire's work in the United States, see "On the Vaccine of Yellow Fever," *Medical News* 51(1887):330a–334a; "The Practibility of Yellow Fever Inoculations as a Prophylactic Measure," *Gaillard's Medical Journal* 47(1889):548–558. On Freire's results, see *Statistiques des vaccinations au moyen de la culture atténuée des microbes de la fièvre jaune* (Berlin, 1891); *Réfutation des recherches sur la fièvre jaune* (Rio de Janeiro, 1888); and numerous articles in *Comptes rendus hebdomadaires de l'Académie des Sciences* 104(1887):1020–1022; 109(1889): 715–716; 113(1891):579–581.

14
G. Sternberg, *Report on the Etiology and Prevention of Yellow Fever* (Washington, 1890), p. 161. See also J. R. Harrison and J. H. Moxley, "Rapport sur certaines expériences . . . ou réfutation des expériences et opinion du Dr. Freire, sur la fièvre jaune," *Le Moniteur scientifique* 15(1885):710.

15
V. Babès, "Sur les microbes trouvés dans le foie et dans le rein d'individus morts de la fièvre jaune," *Comptes rendus hebdomadaires des séances de l'Académie des Sciences* 97(1883):682.

16
A. Le Dantec, *Recherches sur la fièvre jaune* (Paris, 1886), p. 13.

17
Sternberg, *Report,* p. 165. In the United States, see also the critiques of Kinyoun, "Report of Work in the New York Laboratory," *Annual Report of the Supervising Surgeon-General of the Marine Hospital of the United States* (Washington, 1889), p. 105; and of S. Weir Mitchell, "Remarks in Regard to Dr. Finlay's Researches with Reference to the Bacillus of Yellow Fever," *Transactions of the College of Physicians of Philadelphia* 10(1888):65–68. In Cuba, see the critiques of Tamayo, "Los microcoques del Dr. Finlay," *Cronica Médico-Quirurgica de La Habana* 13(1887):439–443, 502–506, 718–720.

18
V. Babès, "Contribution à l'étude des lésions aiguës des reins liées à la présence des microbes. Le rein et le foie dans la fièvre jaune," *Archives de physiologie normale et pathologique* 12(1883):453. See also "Sur les microbes trouvés dans le foie et dans le rein d'individus morts de la fièvre jaune," *Comptes rendus de l'Académie des Sciences* 97(1883):683.

19
P. Gibier, "Yellow Fever. An Experimental Research on Its Etiology," *Medical News* 53(1889):95a. For the discovery made in Havana, see Gibier, "Etude sur l'étiologie de la fièvre jaune," *Comptes Rendus de l'Académie des Sciences* 106(1888):502, and

"Another Yellow-Fever Microbe Found," *Medical Record* 34(1888):183b–184a.

20
Sternberg, *Report,* p. 222.

21
A. Cornil and V. Babès, *Les bactéries* (Paris, 1886), p. 528.

22
Sternberg, *Report,* pp. 167–168.

23
Ibid., p. 222.

24
"Revues et analyses," *Annales de l'Institut Pasteur* 5(1891):796–797.

25
W. Havelburg, "Recherches expérimentales et anatomiques sur la fièvre jaune," *Annales de l'Institut Pasteur,* 1897, p. 521. A longer version of the paper was published in *Berliner Klinische Wochenschrift* 34(1897):493–496, 526–528, 542–544, and 564–567. And following Havelburg was P. Caldas, "Yellow Fever, An Infection Produced by Malignant Colon Bacilli," *Medical News* 75(1899):279a–b.

26
J. Sanarelli, "Premières expériences sur l'emploi du sérum curatif et préventif de la fièvre jaune," *Annales de l'Institut Pasteur* 6(1897):460.

27
J. Sanarelli, "Etiologie et pathogénie de la fièvre jaune," *Annales de l'Institut Pasteur* 6(1897):457.

28
L. Thoinot, "L'étiologie de la fièvre jaune d'après les travaux les plus récents," *Annales d'hygiène publique et de médecine légale* 38(1897):142.

29
F. Novy, "The Etiology of Yellow Fever," *Medical News* 73(1898):368b–369a. The following year Novy showed that exposure of Sanarelli's bacillus to temperatures below minus ten degrees centigrade did not eliminate its virulence: "The resistance of the bacillus icteroides to cold seemed to me incompatible with its supposed role as the cause of yellow fever." "The Bacillus Icteroides: A Reply to Dr. Sanarelli," *Medical News* 75(1899):385b.

30
M. Warner, "Hunting the Yellow Fever Germ: The Principle and Practice of Etiological Proof in Late Nineteenth-Century America," *Bulletin of the History of Medicine* 59(1985):375.

31

G. Sternberg, "Recent Research Relating to the Etiology and Specific Treatment of Yellow Fever," *Medical News* 71(1897):614a. See Sternberg's numerous other articles: "The Bacillus Icteroides of Sanarelli (Bacillus *x*–Sternberg)," *American Journal of the Medical Sciences* 114(1897):307; "Bacillus Icteroides (Sanarelli) and Bacillus X (Sternberg)," *Medical Record* 53(1898):712a–b; "The Bacteriology of Yellow Fever," *Johns Hopkins Hospital Bulletin* 9(1898):119a–120b; "Bacillus Icteroides and Bacillus X," *Journal of the American Medical Association* 30(1898):233b. Concerning the highly favorable reception of Sanarelli's work in the United States, see "Yellow Fever," *Journal of the American Medical Association* 31(1898):832a; R. Wilson, "The Etiology of Yellow Fever," *Yale Medical Journal* 4(1898):221–228; E. Jordan, "Sanarelli's Work upon Yellow Fever," *Science* 6(1897):981b–985a.

32

E. Wasdin, "Yellow Fever: Its Nature and Cause," *Journal of the American Medical Association* 35(1900):871a.

33

W. Reed and J. Carroll, "Bacillus icteroides and Bacillus cholerae suis—A Preliminary Note," 1899, in *Yellow Fever* (Washington, 1911), p. 55. At the same time Sternberg refuted Klebs's discovery in "Dr. Klebs' Amoeba of Yellow Fever," *Journal of the American Medical Association* 30(1898):1054a–1055a.

34

E. Wasdin and H. Geddings, *Report of Commission of Medical Officers* (Washington, 1899), p. 46. The results were confirmed by Archinard, who isolated the bacillus in thirty-two cases out of thirty-nine (see P. Archinard, R. Woodson, and J. Archinard, "Bacteriological Study in the Aetiology of Yellow Fever," *New Orleans Medical Journal* 69[1899]:114a), and by O. Pothier, "Summary of Pathologic and Bacteriologic Work Done at Isolation Hospital, New Orleans, Louisiana," *Journal of the American Medical Association* 30(1898):888a. On corroborative work concerning the toxin, see J. Lacerda and Ramos, "Le bacille ictéroïde et sa toxine (expérience de contrôle)," *Archives de médecine expérimentale et d'anatomie pathologique* 11(1899):394–395. On the question of the diagnosis and serum, see P. Archinard, R. Woodson, and J. Archinard, "The Serum Diagnosis of Yellow Fever Illustrating the Value of Widal's Reaction with the Bacillus Icteroides," *Journal of the American Medical Association* 30(1898):460b–462a; A. Doty, "The Report of a Case Treated with Yellow Fever Serum," *Medical Record* 56(1899):289a–290b. On the controversy, see the articles by Sanarelli, Sternberg, Novy, Reed, and Carroll in *Medical News*

75(1899):193a–199a, 278a–279a, 321a–329b, 385a–388b, 737a–744a, 767a–768a.

35
A. Agramonte, "Report of Bacteriological Investigations upon Yellow Fever," *Medical News* 76(1900):205b.

36
W. Reed, J. Carroll, J. Lazear, A. Agramonte, "The Etiology of Yellow Fever—A Preliminary Note," in *Yellow Fever* (Washington, 1911), p. 69.

5 The Theory Verified

1
P. Manson, "On the Nature and Significance of the Crescentic and Flagellated Bodies in Malarial Blood," *British Medical Journal* 2(1894), in B. H. Kean et al., eds., *Tropical Medicine and Parasitology: Classic Investigations,* 2 vols. (Ithaca, 1978), vol. 1, p. 117a.

2
G. MacCallum "On the Haematozoan Infection of Birds," *Journal of Experimental Medicine* 3(1898), in ibid., p. 65a.

3
"The Role of the Mosquito in the Evolution of the Malaria Parasite," *The Lancet* 2(1898), in ibid., p. 61b.

4
T. Bancroft, "On the Metamorphosis of the Young Forms of *Filaria bancrofti,* Cobb. (*Filaria sanguinis hominis,* Lewis; *Filaria nocturna,* Manson), in the body of *Culex ciliaris,* Linn., the House Mosquito of Australia," *Journal and Proceedings of the Royal Society of New South Wales* 33(1899):51.

5
G. Low, "A Recent Observation on *Filaria nocturna* in *Culex:* Probable Mode of Infection of Man," *British Medical Journal* 1(1900):1457b. James came to the same conclusion in *British Medical Journal* 1(1900):533a. Concerning the history of work on filaria, see R. Blanchard, "Transmission de la filariose par les moustiques," *Archives de Parasitologie* 3(1900):280–291.

6
J. Guiart, "Les moustiques, importance de leur rôle en médecine et en hygiène," *Annales d'hygiène et de médecine légale* 46(1900):435–436. In the United States see C. Craig, "The Transmission of Disease by the Mosquito," *New York Medical Journal* 57(1898):457b; C. Beach, "Insects as Etiological Factors in Disease," *Proceedings of the Connecticut Medical Society,* 1899, p. 110; "Transmission of the Yellow Fever through Mos-

quitoes," *Journal of the American Medical Association* 32(1899):1123a–b.

7
G. Sternberg, "The Transmission of Yellow Fever by the Mosquitoes," in *The Popular Science Monthly,* July 1901, reprinted in his *Sanitary Lesson of the War* (1912), pp. 56–57, and cited by A. Truby, *Memoir of Walter Reed* (New York and London, 1943), p. 90. This version may also be found in L. Howard, *Mosquitoes* (New York, 1901), p. 123, and in J. Peabody, *The Conquest of Yellow Fever* (New York, 1932), p. 6.

8
Testimony of Aristide Agramonte, August 31, 1908, cited by G. Torney and R. Owen, "Yellow Fever Commission," Senate Document No. 520, in *Yellow Fever* (Washington, 1911), p. 25. Agramonte stresses this point in "A Review of Research in Yellow Fever," *Annals of Internal Medicine* 2(1928):146b. This version of events may be found in H. Kelly, *Walter Reed and Yellow Fever* (New York and Baltimore, 1907), p. 142; E. Richter, "Henry R. Carter. An Overlooked Skeptical Epidemiologist," *New England Journal of Medicine* 277(1967):737a; F. Winter, "The Romantic Side of the Conquest of Yellow Fever," *The Military Surgeon* 61(1927):439–440.

9
Torney and Owen, "Yellow Fever Commission," p. 24. The War Department stated: "When the Yellow Fever Commission . . . assembled in Havana, they had no thought of investigating the connection of the mosquito with the spread of yellow fever." See R. O'Reilly, "Experiments Conducted for the Purpose of Coping with the Yellow Fever," Senate Document No. 10, in *Yellow Fever* (Washington, 1911), p. 17. Furthermore, Sternberg later acknowledged that his verbal instructions pertained only to inoculation of blood by means of a syringe. See "Researches Relating to the Etiology of Yellow Fever," *Pan-American Surgical and Medical Journal* 21(1916):20a.

10
Cited in C. Smart, "The Germ of Yellow Fever," *Philadelphia Medical Journal* 6(1900):754b.

11
Truby, *Memoir of Walter Reed,* pp. 89–91. The visit of Finlay took place in early August, not late June or early July. Nevertheless, Gorgas, like Truby, placed the visit in early June, citing Finlay himself as evidence: "Dr. Finlay says on page 1 of his *Agreement Between the History of Yellow Fever and Its Transmission by the Culex Mosquito:* 'The experiments made by Drs. Reed, Carroll, Agramonte, and Lazear were started in June' " (*Sanitation in Panama* [London and New York, 1915], p. 166). It

would have been better to indicate that this was a slip. As for Lazear, it is known that he had been working on malaria in Havana since February 1900. Agramonte remarked: "He [Lazear] had already bred numerous varieties in his laboratory at Camp Columbia." See "To the Editor," *Journal of the American Medical Association* 40(1903):1661a. See also J. del Regato, "Jesse William Lazear," *P and S Quarterly* 16(1971):6a.

12
Winter, "The Romantic Side," pp. 439–440.

13
J. Lazear, "Letters to his Mother and Wife," 1874–1900, cited by del Regato, "Jesse William Lazear," p. 6a.

14
L. Howard, *A History of Applied Entomology* (Washington, 1930), p. 470.

15
H. Durham and W. Myers, "Liverpool School of Tropical Medicine: Yellow Fever Expedition. Some Preliminary Notes," *British Medical Journal* 2(1900):656b. On the period of extrinsic incubation, see H. Carter, "A Note on the Interval Between Infecting and Secondary Cases of Yellow Fever from the Records of the Yellow Fever at Orwood and Taylor, Miss., in 1898," *New Orleans Medical and Surgical Journal* 52(1900):617–636. Manson was the first to suggest that the propagation of yellow fever required an "intermediate host." Oddly enough, he mentioned Finlay's hypothesis but did not relate it to this idea. See *Tropical Diseases* (New York, 1898), p. 128.

16
R. O'Reilly, "Experiments Conducted for the Purpose of Coping with Yellow Fever," Senate Document No. 10, in *Yellow Fever* (Washington, 1911), p. 18. For the description of events I used the documents provided by O'Reilly, G. Torney, and R. Owen, and Agramonte's "The Inside History of a Great Discovery," in A. McGehee Harvey, "Johns Hopkins and Yellow Fever. A Story of Tragedy and Triumph," *Johns Hopkins Medical Journal* 149(1981):35a.

17
J. Carroll, "A Brief Medical Review of the Etiology of Yellow Fever," *New York Medical Journal*, 1904, cited in J. Hemmeter, *Master Minds in Medicine* (New York, 1927), pp. 304–305. Agramonte stated that Carroll had allowed himself to be bitten "in a spirit of derision, without any faith in the mosquito theory." See "Letter from Agramonte," April 12, 1905, in C. Finlay, *OC*, vol. 6, p. 125. This seems probable. We know that Carroll believed that his initial symptoms were the result of a malaria attack.

Nevertheless, Lazear scrupulously recorded this bite along with the rest.

18
Reed et al., "The Etiology," in *Yellow Fever* (Washington, 1911), p. 69.

19
Finlay, *OC*, vol. 6, p. 109.

20
Reed-Carroll correspondence, University of Maryland, cited by T. Woodward, "Yellow Fever: From Colonial Philadelphia and Baltimore to the Mid-Twentieth Century," in A. Lilienfield, ed., *Times, Places, and Persons* (Baltimore and London, 1978), p. 125.

21
W. Reed, "The Propagation of Yellow Fever—Observations Based on Recent Researchers," 1901, in *Yellow Fever* (Washington, 1911), p. 94. On p. 107 Reed indicates that the observation of the two nurses is the only one that corroborates Carter's results.

22
"Letter from Finlay to Delgado," 1901, *OC*, vol. 4, p. 11.

23
"Letter from Reed to Carroll," cited in Bean, *Walter Reed,* p. 135.

24
"Address at the Unveiling of a Tablet to Dr. Jesse William Lazear at the Johns Hopkins Hospital, October 5, 1904," *Johns Hopkins Hospital Bulletin* 15(1904):388b.

25
Letter of W. Reed. to J. Carroll, dated 1 P.M., September 7, 1900, in Woodward, "Yellow Fever," p. 24. His lack of interest in the mosquito theory is well substantiated: "At the time, neither Drs. Reed, Carroll nor myself believed in said theory." See Agramonte, "To the Editor," *Journal of the American Medical Association* 40(1903)1661a. Reed is known to have read Finlay's memoir upon his return from Washington: "I have found among Dr. Lazear's books your copy of the *Annales de la Real. Acad.,* vol. 18, and I am now reading your article." "Letter from Reed to Finlay," October 7, 1900, *OC*, vol. 6, p. 109.

26
Bean, *Walter Reed,* pp. 121–122.

27
D. Gorton, *The History of Medicine,* 1910, vol. 2, p. 443. Following this line leads to portraying Reed as a man who died a

martyr to science: see Hemmeter, *Master Minds in Medicine*, p. 297, and the obituary notice, "Life of Surgeon Walter Reed, U.S.A.—The Discoverer of the Cause of Yellow Fever and Its Means of Prevention," *Virginia Medical Semi-Monthly* 7(1903):450: "There is good reason to believe that Dr. Reed's health was severely shaken by the anxious experiences in investigating the cause and prevention of yellow fever."

28

L. Wood, "Value of Dr. Reed's Work and Expression of Appreciation," in *Yellow Fever* (Washington, 1911), p. 20. Welch threw the weight of his authority into the balance: "I am in a position to know that the credit for original ideas embodied in this work belongs wholly to Maj. Reed" (ibid).

29

"Yellow Fever and Mosquitoes," *British Medical Journal* 2(1900):1391b. See also "The Etiology of Yellow Fever," *Medical News* 77(1900):701b and *Medical Record*, 1900, p. 698a.

30

F. Garrison, *An Introduction to the History of Medicine* (Philadelphia and London, 1917), p. 733.

31

W. Reed, J. Carroll, and A. Agramonte, "The Etiology of Yellow Fever. An Additional Note," 1901, in *Yellow Fever*, p. 87.

32

J. Carroll, "The Transmission of Yellow Fever," *Journal of the American Medical Association* 40(1903):1433a.

33

Reed et al., "The Etiology of Yellow Fever. An Additional Note," p. 82.

34

W. Reed, "Recent Researches Concerning the Etiology, Propagation, and Prevention of Yellow Fever, by the United States Army Commission," in *Yellow Fever* (Washington, 1911), p. 165. On this point see W. Reed and J. Carroll, "The Etiology of Yellow Fever. A Supplementary Note," 1901–1902, in ibid., pp. 149–160.

6 The Analysis of Differences

1

E. Jeanselme and E. Rist, *Précis de pathologie exotique* (Paris, 1909), p. 160. This theme may also be found in E. H. Ackerknecht, *History and Geography of the Most Important Disease* (New York and London, 1965), p. 58, and J. Lopez Sanchez, *La doctrina finlaista* (Havana, 1981), p. 11.

2
N. Stepan, "The Interplay Between Socio-Economic Factors and Medical Science: Yellow Fever Research, Cuba and the United States," *Social Studies of Sciences* 8(1978):412; also p. 397: "Political and economic factors were more important in the lag than any supposed shortcoming in Finlay's science."

3
Ross Danielson, *Cuban Medicine* (New Brunswick, n.d.), p. 80.

4
F. Dominguez Roldan, *Docteur Carlos J. Finlay, son centenaire (1833), sa découverte (1881)* (Paris, 1935), p. 62.

5
C. Rodriguez Exposito, "Finlay," *Cuadernos de historia de la salud pública* 20(1962):90.

6
Stepan, "The Interplay," p. 397.

7
R. O'Reilly, "Experiments Conducted for the Purpose of Coping with Yellow Fever," Senate Document No. 10, in *Yellow Fever* (Washington, 1911), p. 17.

8
Dominguez Roldan, *Docteur Carlos J. Finlay*, p. 117.

9
R. Ross, *Memoirs, with a Full Account of the Great Malaria Problem and Its Solution* (New York, 1923), p. 425. Carroll also stressed this point: "The Transmission of Yellow Fever" *Journal of the American Medical Association* 40(1903):1431b–1432a.

10
Dominguez Roldan, *Docteur Carlos J. Finlay*, p. 141. Stepan, "The Interplay," reveals the same confusion: "Finlay proposed in 1881 that *Aedes aegypti* mosquito was the intermediary host of yellow fever."

11
Rodriguez Exposito, "Finlay," p. 100.

12
C. Finlay and C. Delgado, "Estado actual de nuestros conocimientos tocantes a la fiebre amarilla," 1887, OC, vol. 1, p. 443. King, who thoroughly understood Finlay's hypothesis, wrote in "Insects and Disease—Mosquitoes and Malaria," *Popular Science Monthly* 23(1883):655: "What product of man's art has not been anticipated by nature?"

13
L. J. B. Béranger-Féraud, *Traité théorique et clinique de la fièvre jaune* (Paris, 1890), p. 593. This initial improbability was com-

pounded by an even more incredible concatenation of vehicles: the mosquito in an article of clothing, clothing in a trunk, and the trunk on board a ship.

14
Ibid., p. 591. Corre had formulated the same criticism as early as 1883. See "Sur une nouvelle théorie pathogénique de la fièvre jaune," *Archives de médecine navale* 39(1883):69–70.

15
See C. Finlay, "Inoculación de la fiebre amarilla mediante mosquitos contaminados," 1891, *OC,* vol.2, pp. 95–98, and "Inmunidad a la fiebre amarilla. Formas de propagación. Teoría del mosquito," 1894, ibid., pp. 125–130.

16
G. Sternberg, "Dr. Finlay's Mosquito Inoculations," *American Journal of the Medical Sciences* 102(1891):627–628.

17
C. Finlay, "Dos maneras distinctas de transmitirse la fiebre amarilla por el *Culex Mosquito (Stegomyia Taeniata),*" 1901, *OC,* vol. 3, p. 50. Finlay repeatedly stated that the mosquito could transmit yellow fever during the first three days. See "Piezas constitutivas de la trompa del *Culex Mosquito,*" 1902, ibid., pp. 58–59; "Del mosquito como factor etiologico de la fiebre amarilla," 1906, ibid., p. 327.

18
R. Ross, *La découverte de la transmission du paludisme par les moustiques* (Paris, 1929), p. 45.

19
"Letter from Reed to his wife," in Sir R. Boyce, *Mosquito or Man?* (London, 1911), p. 131.

20
"The Etiology of Yellow Fever," *Medical News* 77(1900):701a.

21
G. Canguilhem, *Le normal et le pathologique* (Paris, 1966), p. 11.

22
F. Courtès, "Georges Cuvier ou l'origine de la négation," *Thalès* 13(1969):12.

23
G. Canguilhem, *Etudes d'histoire et de philosophie des sciences* (Paris, 1976), p. 149.

24
K. Goldstein, *La structure du comportement* (Paris, 1963), p. 313.

25
R. Leuckart, *The Parasite of Man* (Edinburgh, 1886), p. 64, note

164   2.

26
G. Sternberg, "Yellow Fever," in *A System of Practical Medicine* (London, 1897), vol. 1, p. 267.

27
"Patrick Manson's Letter to Finlay," 1909, *OC,* vol. 6, p. 134.

7 The Spoils of Discovery

1
Quoted by Dominguez Roldan, *Docteur Carlos J. Finlay,* p. 132.

2
"Letter from Finlay to Delgado," January 2, 1901, *OC,* vol. 6, p. 13.

3
Quoted by Bean, *Walter Reed,* p. 178.

4
Reed, "The Propagation," p. 95.

5
"Letter from Finlay to Manson," September 1908, *OC,* vol. 6, p. 129. Elsewhere we find the same admission: "I had not reached the more advanced views concerning the development of the germ in the salivary glands of the infected mosquito." See C. Finlay, "Yellow Fever and Its Transmission by Means of the *Culex Mosquito,*" 1886, *OC,* vol. 1, p. 427.

6
C. Finlay, "La fiebre amarilla, su transmisión por intermedio del *Culex mosquito,*" 1886, *OC,* vol. 1, p. 413. Finlay had expressed the same idea two years earlier: "Fiebre amarilla experimental comparada con la natural en sus formas benignas," 1884, ibid., pp. 336–337. As we saw earlier, this supposition enabled him to overcome the objection raised by Corre at that very moment, in 1883.

7
C. Finlay, "Transmisión de la fiebre amarilla por el *Culex mosquito,*" 1891 manuscript, *OC,* vol. 3, p. 174. See also "Factores climatológicos concernientes a la propagación de la fiebre amarilla en La Habana," 1893, ibid., p. 130. From this came the remarkable explanation of the Saint Nazaire epidemic. On this point there was much confusion. See the very detailed study of W. Coleman, *Yellow Fever in the North* (Madison, 1987), pp. 79–138. At the time Hammond was one of the few physicians to adopt Finlay's ideas. See "For What Purpose Were Mosquitoes Created?" *Science* 8(1886):436a.

8

C. Finlay, "Método para extirpar la fiebre amarilla recomendado desde 1899," 1902, *OC*, vol. 3, p. 82. Wheat rust is caused by a parasitic fungus whose spores, disseminated by the wind, germinate only on barberry, whose spores in turn germinate only on wheat. See Bary, "Nouvelles observations sur les urédinées," *Annales des Sciences Naturelles, Botanique* 5(1866):262–74. Bary's discovery corroborated the belief of farmers that wheat was liable to attack by rust only when grown in the neighborhood of barberry. This led to the idea of stamping out barberry altogether. See F. Brooks, *Plant Diseases* (London, 1928), p. 237.

9

"Session de l'Académie des Sciences de La Havane," March 12, 1882, *OC*, vol. 4, p. 330. See also C. Finlay, "Consideraciones acerca de algunos casos de Filariasis observados en La Habana," 1882, ibid., p. 22. Van Tieghem of course never discussed an "intermediate host"; he spoke only of a two-part developmental cycle involving a change of host. See P. Van Tieghem, *Traité de botanique* (Paris, 1884), p. 1038. Finlay did indeed read Van Tieghem *in 1886,* but at that time what he took from him was something quite different: "I endeavored to ascertain for that insect the five temperature-limits which Van Tieghem, in his *Traité de Botanique* (Paris, 1884, p. 88) considers *critical* for seeds and plants." See C. Finlay, "Atmospheric Temperature as an Essential Factor in the Propagation of Yellow Fever," 1907, *OC*, vol. 3, p. 349.

10

C. Finlay, "Teoría del mosquito de Finlay antes y después de la investigación oficial sobre la misma," 1901, *OC*, vol. 3, pp. 35–36. The same assertion may be found in "Reseña de los progresos realizados en el siglo XIX en el estudio de la propagación de la fiebre amarilla," 1901, ibid., p. 12, and in "Dos maneras distintas de transmitirse la fiebre amarilla por el *Culex Mosquito (Stegomyia Taeniata)*," ibid., p. 47. A quick glance at the context shows, however, that once again Finlay's version is simply not credible. In November 1889 Finlay performed two operations. First, he drew an analogy between malaria and cholera and typhoid fever with respect to the recently discovered mode of propagation: "These insects [flies] might easily pick up hematozoa in extravasated blood from the excretions of victims . . . and then deposit them in food or beverages." Second, Finlay drew an analogy between yellow fever and Texas fever, again with respect to the mode of propagation: "Once the mosquito is contaminated, we may assume that the pathogenic germs invade not only the eggs but also the veneno-salivary glands, subsequent to which they are injected along with the secretions of those glands along the path of the puncture." See C. Finlay, "Los mosquitos considerados como agentes de la transmisión de la fiebre amarilla y de la malaria," 1898, OC, vol. 2, p. 252. Note, by the way, that the first conjecture was taken from Sternberg: see "Yellow Fever," in A. Loomis and W. Thomson, *A System of Practical Medicine* (London, 1897), vol. 1, p. 271. And the second was taken from Koch, whose lecture, delivered upon his return from Africa, was translated by Finlay: see "Experiencia médica de Robert Koch en los trópicos," 1898, OC, vol. 4, pp. 199–203. Although Finlay imagined and described a process in the same terms as Ross, he never produced a concept of it. Furthermore, the assumption that the malarial hematozoon is transmitted mechanically shows that Finlay had no idea about the life cycles of parasites with alternating hosts. In addition, the assumption that the yellow fever germs pass through the eggs necessarily leads to the idea that the veneno-salivary glands must be the germs' exit route in mosquitoes born from infected eggs. Hence the hypothesis that the germs might be expelled by the veneno-salivary glands of the mosquito during biting was merely an extension of this logical inference.

11
Reed, "The Propagation," p. 96.

12
Ibid., pp. 92–93.

13
"Letter of February 26, 1902," National Library of Medicine, quoted by Richter, "Henry R. Carter," p. 734b.

14
Reed, "Recent Researches," p. 166.

15
Dominguez Roldan, *Docteur Carlos J. Finlay,* p. 127. The same claim is made by most Cuban historians. Compare, for instance, Lopez Sanchez, *La doctrina finlaísta,* p. 28: "Finlay's conclusions were essentially the same as those of the United States Military Medical Commission."

16
Rodriguez Exposito, "Finlay," *Cuadernos de historia de la salud pública* 20(1962):110.

17
J. Guiteras, "Dr. Carlos Finlay. Biographical Notes," OC, vol. 1, pp. 14–15.

18
Dominguez Roldan, *Docteur Carlos J. Finlay,* pp. 30–31. This is the explanation preferred by Cuban historians. See Lopez San-

chez, *La doctrina finlaísta,* pp. 12–13, 26–27, 38; and S. Amaro Mendez, *Alas amarillas* (Havana, 1983), pp. 44–45.

19
Chauncey D. Leake, "Stories of the Men Who Conquered Yellow Fever," in A. Smith, *Yellow Fever in Galveston 1839* (Austin, 1951), pp. 126–127. For the sake of completeness, it should be noted that Carroll, Agramonte, and Ross stated that Finlay was familiar with Manson's work. They gave no proof; more than that, their statements were made for the sole purpose of diminishing Finlay's achievement. See J. Carroll, "A Brief Review of the Aetiology of Yellow Fever," *The New York Medical Journal* 79(1904):307a; A. Agramonte, "An Account of Dr. Louis Daniel Beauperthuy, A Pioneer in Yellow Fever Research," *Boston Medical and Surgical Journal* 158(1908):928a; Ross, *Memoirs,* p. 123.

20
G. Nuttall, "In Memoriam. Walter Reed," *The Journal of Hygiene* 3(1903):294.

21
S. Adams, "Memorial Meeting of the Medical Society of the District of Columbia, held on December 31, 1902," in *Yellow Fever* (Washington, 1911), p. 34.

22
M. Gorgas and B. Hendrick, *William Crawford Gorgas, His Life and Work* (Philadelphia and New York, n.d.), pp. 100–101.

23
D. Beauperthuy, "Recherches sur la cause du choléra asiatique, sur celle du typhus ictéroide et des fièvres marécageuses," *Comptes Rendus hebdomadaires des séances de l'Académie des Sciences* 42(1856):692.

24
"Enfermedad de los cocoteros," session of March 22, 1882, *OC,* vol. 4, p. 15. No one has noticed that in 1889 the problem would be posed in the following terms: "Ramos's *uredo cocivore* must undergo a transformation similar to *Puccinia graminis* in wheat, since it is also a heteroecious fungus. It is therefore of the utmost importance to determine what plant plays the role of the barberry in the coconut disease . . . because stamping it out might be a way of halting the epidemic." See D. Tamayo, "La enfermedad de los cocoteros," *Cronica Médico-Quirurgica de La Habana* 15(1889):485. It seems likely that Finlay, after the fact, used this article by Tamayo to prove that he had envisioned the notion of an "intermediate host" and suggested a means of eradicating yellow fever. Historians, moreover, have not even been as cautious

as Finlay, who limited himself to saying that this analogy led him to the idea of the mosquito *as vehicle*. Strictly speaking, the agent of transmission of *Puccinia graminis* is the wind. Hence if Finlay had considered this analogy, he would have rejected it for the same reason he rejected the nidus theory. Recently, J. Lopez Sanchez, *Finlay, el hombre y la verdad científica* (Havana, 1987), pp. 154–157, has tried to resolve the question by arguing that in 1878 Finlay read the 1874 French translation of Sachs's *Treatise*. Unfortunately, in the 1874 edition van Tieghem makes no allusion to the practical consequences of understanding the *Puccinia graminis* cycle. In Cuba Finlay read the 1874 edition, but the historians cite the 1884 edition.

25
L. Wood, quoted in Dominguez Roldan, *Docteur Carlos J. Finlay,* p. 11. In 1943 Sol Bloom delivered an enthusiastic speech to the House of Representatives in which he recognized Finlay's priority and explained why Finlay was so little known in the United States: the brilliant corroboration of his theory and spectacular eradication of the disease allegedly overshadowed the ingenious Cuban's austere works. See Sol Bloom, "Dr. Carlos J. Finlay," *Booklet on Sanitation History* 13(1959):10.

26
G. Leader, "Día centenario del Dr. Carlos Finlay," *Cuadernos de Historia Sanitaria* 15(1959):94.

27
L. Howard, *A Half Century of Public Health* (New York, 1921), p. 421.

28
H. Portell Vila, *Historia de Cuba en sus relaciones con los Estados Unidos y España* (Havana, 1941), p. 105.

29
"The First Battle against Yellow Fever," *Medical Record* 54(1898):126b.

30
P. Ashburn, *A History of the Medical Department of the United States Army* (Boston and New York, 1929), p. 263.

31
Hemmeter, *Master Minds in Medicine,* p. 311.

32
S. Fernandez, "Homenaje Finlay," in *Sanidad y Beneficiencia* 20(1918):15. Dominguez Roldan envisioned an even more splendid monument to be built in Panama overlooking the canal: "America should place a statue of Finlay at the highest point of the Culebra, so that it might contemplate the grand work that

was the result of his brilliant discovery." See *Docteur Carlos J. Finlay,* p. 263.

33
C. Rodriguez Exposito, "Finlay: polémica permanente," *Cuadernos de Historia de la Salud Pública* 17(1956):18.

34
H. Kelly and W. Burrage, *American Medical Biographies* (Baltimore, 1920) p. 965a.

35
C. Rodriguez Exposito, "En homenaje a Finlay," in "Trabajos Académicos y otros estudios," *Cuadernos de Historia de la Salud Pública* 70(1985):204.

36
J. Guiteras, "Dr. Osler's Address on 'The Nation and the Tropics and Dr. Finlay,' " *Sanidad y beneficiencia* 20(1918):124.

37
E. Roig de Leuchseuring, *Médicos y Medicinas en Cuba* (Havana, 1965), pp. 139–140.

38
H. Abascal and C. Rodriguez Exposito, "Permanencia de la doctrina de Finlay ante el XV Congreso Internacional de Historia de la Medicina," *Cuadernos de historia sanitaria* 11(1957):51. In 1971 a fourth petition was filed. This time the Cubans felt threatened by Rosario Beauperthuy de Benedetti, who claimed priority for Beauperthuy. See C. Rodriguez Exposito, "Finlay por cuarta vez ante el Congreso Internacional de Historia de la Medicina," *Cuadernos de Historia de la salud pública* 52(1971).

39
H. Harant, *Histoire de la parasitologie* (Paris, 1955), p. 26.

40
F. Delaporte, *Disease and Civilization* (Cambridge, Mass., 1986).

41
G. Deleuze and F. Guattari, *Rhizome* (Paris, 1976), p. 37.

Bibliography

PERIODICALS (AMERICAN, BRITISH, CUBAN, FRENCH)

American Journal of the Medical Sciences
Annales de l'Institut Pasteur
Annales des Sciences Naturelles
Annales d'hygiène publique et de médecine légale
Annals of Internal Medicine
Annual Report of the National Board of Health
Annual Report of the Supervising Surgeon-General of the Marine-Hospital of the United States
Archives de médecine expérimentale et d'anatomie pathologique
Archives de médecine navale
Archives de parasitologie
Archives de physiologie normale et pathologique
Atlanta Medical and Surgical Journal
Booklet on Sanitation History
Boston Medical and Surgical Journal
British Medical Journal
Bulletin de l'Académie de Médecine
Bulletin of the History of Medicine
Comptes rendus de l'Académie des Sciences
Comptes rendus hebdomadaires de l'Académie des Sciences
Cronica Médico-Quirurgica de La Habana
Cuadernos de historia de la salud pública

172   *Gaillard's Medical Journal*

Gazette des hôpitaux

International Review of Tropical Medicine

Johns Hopkins Hospital Bulletin

Johns Hopkins Medical Journal

Journal and Proceedings of the Royal Society of New South Wales

Journal of Experimental Medicine

Journal of Hygiene

Journal of the American Medical Association

Journal of the Linnean Society

Journal of the Medical Association

Journal of Tropical Medicine and Hygiene

Lancet

Medical News

Medical Record

Medical Reports. China Imperial Maritime Customs

Military Surgeon

Moniteur scientifique

New England Journal of Medicine

New Orleans Medical and Surgical Journal

New York Medical Journal

Pan-American Surgical and Medical Journal

P and S Quarterly

Philadelphia Medical Journal

Popular Science Monthly

Proceedings of the Connecticut Medical Society

Revue d'hygiène et de police sanitaire

Sanidad y Beneficiencia

Sanitarian

Science

Social Studies in Science

Thalès

Transactions of the College of Physicians of Philadelphia

Transactions of the Linnean Society
Transactions of the Pathological Society
Virginia Medical Semi-Monthly
Yale Medical Journal

PRIMARY AND SECONDARY SOURCES

Ackerknecht, E. H. *History and Geography of the Most Important Disease.* New York and London, 1965.

Amaro Mendez, S. *Alas amarillas.* Havana, 1983.

Ashburn, P. *A History of the Medical Department of the United States Army.* Boston and New York, 1929.

Bean, W. *Walter Reed. A Biography.* Charlottesville, 1982.

Béranger-Féraud, J.-B. *Traité théorique et clinique de la fièvre jaune.* Paris, 1890.

Blanchard, R. *Les moustiques, histoire naturelle et médicale.* Paris, 1905.

Carmona y Valle, M. *Leçons sur l'étiologie et la prophylaxie de la fièvre jaune.* Mexico City, 1885.

Chaillé, S. *Conclusions of the Board of Experts Authorized by Congress to Investigate the Yellow Fever Epidemic of 1878.* Washington, 1879.

Cornil, A., and V. Babès. *Les bactéries.* Paris, 1886.

Davaine, C. *Traité des entozoaires.* Paris, 1877.

Dean, R. *Report on Yellow Fever in the U.S.S. Plymouth in 1878'9.* Washington, 1880.

Dominguez Roldan, F. *Docteur Carlos J. Finlay, son centenaire (1833), sa découverte (1881).* Paris, 1935.

Ffirth, S. *A Treatise on Malignant Fever with an Attempt to Prove Its Non-Contagious Nature.* Philadelphia, 1804.

Finlay, C. *Obras Completas.* 6 vols., Havana, 1965–1983.

Freire, D. *Doctrine microbienne de la fièvre jaune et ses inoculations préventives.* Rio de Janeiro, 1885.

Garrison, F. *An Introduction to the History of Medicine.* Philadelphia and London, 1917.

Gorgas, M., and B. Hendrick. *William Crawford Gorgas, His Life and Work.* Philadelphia and New York, 1943.

Gorgas, W. *Sanitation in Panama*. London and New York, 1915.

Hemmeter, J. *Master Minds in Medicine*. New York, 1927.

Howard, L. *A Half Century of Public Health*. New York, 1921.

Howard, L. *A History of Applied Entomology*. Washington, 1930.

Howard, L. *Mosquitoes*. New York, 1901.

Jeanselme, E., and E. Rist. *Précis de pathologie exotique*. Paris, 1909.

Kelly, H. *Walter Reed and Yellow Fever*. New York and Philadelphia, 1907.

Kelly, H., and W. Burrage. *American Medical Biographies*. Baltimore, 1920.

La Roche, R. *Yellow Fever, Considered in Its Historical, Pathological, Etiological, and Therapeutical Relations*. 2 vols., Philadelphia, 1853.

Le Dantec, A. *Recherches sur la fièvre jaune*. Paris, 1886.

Leuckart, R. *The Parasite of Man*. Edinburgh, 1886.

Loomis, A., and W. Thomson. *A System of Practical Medicine*. London, 1897.

Lopez Sanchez, J. *La doctrina finlaista*. Havana, 1981.

Manson, P. *Tropical Diseases*. New York, 1898.

Manson-Bahr, P., and A. Alcock. *The Life and Work of Sir Patrick Manson*. London, 1927.

Peabody, J. *The Conquest of Yellow Fever*. New York, 1932.

Portell Vila, H. *Historia de Cuba en sus relaciones con los Estados Unidos y España*. Havana, 1941.

Reed, W., J. Carroll, J. Lazear, and A. Agramonte. Reports and memoirs in *Yellow Fever*. Washington, 1911.

Roig de Leuchseuring, E. *Médicos y Medicinas en Cuba*. Havana, 1965.

Ross, D. *Cuban Medicine*. New Brunswick, n.d.

Ross, R. *La découverte de la transmission du paludisme par les moustiques*. Paris, 1929.

Ross, R. *Memoirs, with a Full Account of the Great Malaria Problem and Its Solution*. New York, 1923.

Scott, H. *A History of Tropical Medicine*. 3 vols., Baltimore, 1939.

Singer, C., and E. Ashworth Underwood. *A Short History of Medicine*. New York and Oxford, 1939.

Sternberg, G. *Report on the Etiology and Prevention of Yellow Fever*. Washington, 1890.

Sternberg, G. *Sanitary Lesson of the War*. 1912.

Truby, A. *Memoir of Walter Reed*. New York and London, 1943.

Van Tieghem, P. *Traité de botanique*. Paris, 1884.

Wasdin, E., and H. Geddings. *Report of Commission of Medical Officers*. Washington, 1899.

Biographical Index

DATE DUE

		NOV 1 1 2014	